Pearson Revise

Pearson Edexcel GCSE (9–1)
Mathematics
Foundation tier
Revision Workbook

Series Consultant: Harry Smith

Author: Navtej Marwaha

Get the inside track

Look out for these features to help turbo-charge your revision:

 These questions cover skills and techniques that real students have struggled with in recent exams. Check out the corresponding *Revision Guide* page for more top tips and things to watch out for.

 You will have to use problem-solving skills throughout your exam. Boxes with this icon will highlight problem-solving skills and strategies to help you stay ahead of the pack.

 We've picked 25 of the hottest topics. These pages contain key skills and knowledge that you're likely to need in your upcoming exams. If you're pushed for time you might want to practise these first.

 There is some tough material in GCSE Maths. We've identified 25 of the trickiest topics. You might want to save these topics for days when you have a bit more time to concentrate on them.

 Where you see this icon, part of the answer has been completed for you.

 This scale tells you how difficult each question is.

A small bit of small print

Pearson Edexcel publishes Sample Assessment Material and the Specification on its website. This is the official content and this book should be used in conjunction with it. The questions have been written to help you practise every topic in the book. Remember: the real exam questions may not look like this.

Contents

1-to-1 page match with the Revision Guide ISBN 9781447988045

- i Get the inside track
- ii Contents

NUMBER
- 1 Place value
- 2 Negative numbers
- 🦅 3 Rounding numbers
- 4 Adding and subtracting
- 5 Multiplying and dividing
- 6 Decimals and place value
- 7 Operations on decimals
- 8 Squares, cubes and roots
- 9 Indices
- 🦅 10 Estimation
- 11 Factors, multiples and primes
- 12 HCF and LCM
- 🦅 13 Fractions
- 14 Operations on fractions
- ⚠ 15 Mixed numbers
- 🦅 16 Calculator and number skills
- 🦅 17 Standard form 1
- 18 Standard form 2
- 19 Counting strategies
- 20 Problem-solving practice 1
- 21 Problem-solving practice 2

ALGEBRA
- 🦅 22 Collecting like terms
- 23 Simplifying expressions
- 24 Algebraic indices
- 🦅 25 Substitution
- 26 Formulae
- 27 Writing formulae
- 28 Expanding brackets
- 29 Factorising
- 30 Linear equations 1
- ⚠ 31 Linear equations 2
- 32 Inequalities
- 33 Solving inequalities
- 34 Sequences 1
- 🦅 35 Sequences 2
- 36 Coordinates
- 37 Gradients of lines
- 🦅 38 Straight-line graphs 1
- ⚠ 39 Straight-line graphs 2
- 40 Real-life graphs
- 41 Distance–time graphs
- 42 Rates of change
- ⚠ 43 Expanding double brackets
- 44 Quadratic graphs
- 45 Using quadratic graphs
- ⚠ 46 Factorising quadratics
- ⚠ 47 Quadratic equations
- ⚠ 48 Cubic and reciprocal graphs
- ⚠ 49 Simultaneous equations
- ⚠ 50 Rearranging formulae
- 🦅 51 Using algebra
- ⚠ 52 Identities and proof
- 53 Problem-solving practice 1
- 54 Problem-solving practice 2

RATIO & PROPORTION
- 55 Percentages
- 🦅 56 Fractions, decimals and percentages
- 57 Percentage change 1
- 58 Percentage change 2
- 🦅 59 Ratio 1
- 60 Ratio 2
- 🦅 61 Metric units
- ⚠ 62 Reverse percentages
- ⚠ 63 Growth and decay
- 64 Speed
- 65 Density
- 66 Other compound measures
- 67 Proportion
- ⚠ 68 Proportion and graphs
- 69 Problem-solving practice 1
- 70 Problem-solving practice 2

GEOMETRY & MEASURES
- 71 Symmetry
- 72 Quadrilaterals
- 🦅 73 Angles 1
- 74 Angles 2
- 75 Solving angle problems
- ⚠ 76 Angles in polygons
- 77 Time and timetables
- 78 Reading scales
- 79 Perimeter and area
- 🦅 80 Area formulae
- 81 Solving area problems
- 82 3-D shapes
- 83 Volumes of cuboids
- 84 Prisms
- ⚠ 85 Units of area and volume
- 86 Translations
- 87 Reflections
- 88 Rotations
- 89 Enlargements
- 🦅 90 Pythagoras' theorem
- ⚠ 91 Line segments
- ⚠ 92 Trigonometry 1
- ⚠ 93 Trigonometry 2
- ⚠ 94 Exact trigonometry values
- 95 Measuring and drawing angles
- 96 Measuring lines
- 🦅 97 Plans and elevations
- 98 Scale drawings and maps
- 99 Constructions 1
- 100 Constructions 2
- 101 Loci
- 102 Bearings
- 103 Circles
- 🦅 104 Area of a circle
- ⚠ 105 Sectors of circles
- 106 Cylinders
- 107 Volumes of 3-D shapes
- ⚠ 108 Surface area
- 109 Similarity and congruence
- 🦅 110 Similar shapes
- ⚠ 111 Congruent triangles
- ⚠ 112 Vectors
- 113 Problem-solving practice 1
- 114 Problem-solving practice 2

PROBABILITY & STATISTICS
- 115 Two-way tables
- 116 Pictograms
- 117 Bar charts
- 🦅 118 Pie charts
- 119 Scatter graphs
- 🦅 120 Averages and range
- 121 Averages from tables 1
- ⚠ 122 Averages from tables 2
- 123 Line graphs
- 124 Stem-and-leaf diagrams
- 125 Sampling
- 126 Comparing data
- 🦅 127 Probability 1
- 128 Probability 2
- 129 Relative frequency
- 130 Frequency and outcomes
- 131 Venn diagrams
- 🦅 132 Set notation
- ⚠ 133 Independent events
- 134 Problem-solving practice 1
- 135 Problem-solving practice 2

136 Paper 1 Practice exam paper

143 Answers

KEY

🦅 = Hot Topic   ⚠ = Tricky Topic

Had a go ☐   Nearly there ☐   Nailed it! ☐   **NUMBER**

Place value

Target grade 1

Guided

1 (a) Write the number nine thousand, three hundred and fifty-one in figures.

9 .... **(1 mark)**

(b) Write the number 4196 in words.

Four thousand, one hundred and .... **(1 mark)**

(c) Write down the value of the 5 in the number 95 872.

5 .... **(1 mark)**

Target grade 1

Guided

2 Write down the number twelve thousand and sixty in a place value table.

....	....	Hundreds	....	Units
....	....	....	6	0

(2 marks)

Target grade 1

3 Write these numbers in order:

(a) 165, 146, 127, 49, 169

Start with the lowest number.

.... **(1 mark)**

(b) 7429, 7249, 7942, 7924, 7028

.... **(1 mark)**

Target grade 1

4 Write these amounts in order:

(a) £63 452, £63 593, £65 601, £63 004, £62 400

.... **(1 mark)**

(b) £1.20, 63p, £1.02, 36p, £1.12   *Convert the pounds into pence and then put the numbers in order.*

.... **(1 mark)**

Target grade 1

5 Peter wrote down his weekly pocket money in order.
£1.80, £1.95, £2.10, £2.01, £2.45, £2.50
Is he correct?

.... **(1 mark)**

Target grade 1

6 Anton is buying supplies for a charity event.
A pack of 50 paper cups costs £1.89.
A pack of 10 paper plates costs 49p.
Anton has £15 to spend.
Anton buys 250 paper cups and spends the rest on paper plates.
How many paper plates can he buy?

Examiners' report: Read questions carefully and copy any figures from the question accurately. Your final answer should be the number of paper plates Anton can buy, not an amount of money.

.... **(1 mark)**

1

NUMBER

Had a go ☐ Nearly there ☐ Nailed it! ☐

Negative numbers

Target grade 1 **Guided**

1 (a) Write the following numbers in order:

   6   −11   −4   0   4

   Start with the lowest number.

   −11   ....   **(1 mark)**

   (b) Work out

   − − = +

   (i) −9 + 7 = ... **(1 mark)**   (ii) −7 − −4 = ... = ... **(1 mark)**

   (iii) −6 − 4 = ... **(1 mark)**   (iv) −10 − +6 = ... **(1 mark)**

Target grade 1 **Guided**

2 (a) Work out

   + × + = +   + × − = −   − × + = −   − × − = +

   (i) −7 × 2 = −14 **(1 mark)**   (ii) 63 ÷ −9 = ... **(1 mark)**

   (iii) −6 × −4 = ... **(1 mark)**   (iv) −42 ÷ −6 = ... **(1 mark)**

   + ÷ + = +   + ÷ − = −   − ÷ + = −   − ÷ − = +

Target grade 1

3 On a certain day in Moscow the temperature at 12 noon is 7 °C but by 6 pm it has dropped by 9 °C. By 9 pm it has dropped a further 5 °C and by 12 midnight it has dropped a further 8 °C. Find the temperature at

   (a) 6 pm .... °C **(1 mark)**   (b) 9 pm .... °C **(1 mark)**   (c) 12 midnight .... °C. **(1 mark)**

Guided

   (d) What was the overall drop in temperature from 12 noon to 12 midnight?

   Temperature at 12 noon = 7 °C

   Temperature at 12 midnight = ...

   Drop in temperature = 7 − .... = ... °C   **(1 mark)**

Target grade 1

4 The table gives information about the highest and lowest temperatures in five cities during one year.

	London	New York	Moscow	New Delhi	Lisbon
Highest temperature (°C)	30	28	25	40	29
Lowest temperature (°C)	−8	−10	−15	−7	−3

   (a) What is the difference between the highest temperature in New Delhi and the lowest temperature in New York?

   Write down 'highest temperature of New Delhi − lowest temperature of New York'. Remember to include units in your answer.

   .... °C   **(1 mark)**

   (b) Which city recorded the biggest difference between the highest and lowest temperature?

   ....   **(1 mark)**

   Viktor works out that the temperature halfway between the lowest temperature in Lisbon and the lowest temperature in Moscow is −10 °C.

   (c) Is Viktor correct? Give a reason for your answer.

   ....   **(1 mark)**

Had a go ☐   Nearly there ☐   Nailed it! ☐

Rounding numbers

Target grade 1 **Guided**

1 Round

(a) 26 723 to the nearest thousand

(b) 6453 to the nearest hundred

(c) 87 536 to the nearest ten.

27 000   **(1 mark)**   6........................ **(1 mark)**   87 5.................... **(1 mark)**

Target grade 3

2 Round 8.635 21 correct to

(a) 1 significant figure

(b) 2 significant figures

(c) 3 significant figures.

.... **(1 mark)**   .... **(1 mark)**   .... **(1 mark)**

Target grade 3 **Guided**

3 Round 0.003 467 2 correct to

(a) 1 significant figure

(b) 2 significant figures

(c) 3 significant figures.

0.003   **(1 mark)**   .... **(1 mark)**   .... **(1 mark)**

> The first significant figure is the number 3.

Target grade 3 **Guided**

4 Round 38 652 correct to

(a) 1 significant figure

(b) 2 significant figures

(c) 3 significant figures.

40 000   **(1 mark)**   .... **(1 mark)**   .... **(1 mark)**

> You need to include enough zeros to show the correct place value.

Target grade 3

5 In her science class, Anjali measured the mass of some objects made from different types of materials. Here are her results.

Material	Wood	Plastic	Metal	Rubber
Mass *m* (g)	20.356	265.800	168.240	127.500

Write down the mass of the

(a) wood to 3 significant figures

(b) plastic to the nearest hundred grams

.... **(1 mark)**   .... **(1 mark)**

(c) metal to 2 significant figures

(d) rubber to the nearest ten grams.

.... **(1 mark)**   .... **(1 mark)**

Target grade 3

6 Jason is weighing some objects on an electronic scale.

0.02346

He writes the answer as 0.023 g correct to 3 significant figures.
Is he correct? Explain your answer.

> **Problem solved!** State whether Jason is correct **and** write some words explaining why. You could write the correct answer, or explain what Jason has done wrong.
>
> Which number is the first significant figure?

.... **(1 mark)**

3

NUMBER

Had a go ☐ Nearly there ☐ Nailed it! ☐

Adding and subtracting

Target grade 1

Guided

1. Work out

   (a) 842 + 158 + 23

   ```
     842
     158
   +  23
   ------
   ......3
   ```

   (1 mark)

   (b) 741 − 164

   ```
    ⁶7 ¹³4̸ ¹1̸
   − 1 6 4
   ---------
        ......7
   ```

   (1 mark)

Target grade 1

2. Work out

   (a) 7263 + 915

   (b) 7629 − 7452

   .... **(1 mark)**   .... **(1 mark)**

Target grade 1

Guided

3. Kevin buys some items from a shop.
   He buys a box of chocolates costing £3.65 and three rolls of wrapping paper costing £1.65 each.
   He gives the cashier a £20 note.
   How much change should he receive?

   Convert the pounds into pence.

   365 + 165 + 165 + 165 = ...

   2000 − .... = ...

   (3 marks)

Target grade 1

4. There are 52 children on the pirate ship at a fairground.
   When the pirate ship stops, 39 children get off and 28 children get on.
   How many children are now on the pirate ship?

   In this case 'get off' means subtract and 'get on' means add.

   You have to show your working. Do not just write down a number.

   .... **(2 marks)**

Target grade 1

5. Part of a receipt is missing.
   David pays £5 and receives 50p change.
   David works out that the coffee cost £2.49.

Slice of cake	95p
Mug of tea	£1.49
Cup of coffee	

   Problem solved! The easiest way to work out whether David is correct is to calculate the cost of a cup of coffee. Remember to show your working and write a conclusion to answer the question.

   Is he correct? Explain your answer.

   ....

   .... **(3 marks)**

Had a go ☐ Nearly there ☐ Nailed it! ☐   **NUMBER**

Multiplying and dividing

1 Work out

(a) 83×23

```
    83
  × 23
  ....
  ....0
  ____
  ....
```

Work out 83×3

Work out 83×2

(b) $972 \div 4$

$$4\overline{)9^17 2}$$

$4 \times 2 = 8$ so 4 divides into 9 twice with remainder 1.

(1 mark) (2 marks)

2 Tins of biscuits come in three sizes. There are 28 biscuits in the small size and four times as many in the medium size. In the large size there are seven times as many as in the small size. How many biscuits are in the

(a) medium size

(b) large size?

.... (1 mark) .... (1 mark)

3 A shop sold 42 boxes of flowers. Each box contained 18 flowers. Work out the total number of flowers sold.

 Show all of your working, writing figures neatly so they can be easily read.

.... (3 marks)

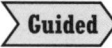

4 Work out

(a) 962×45

```
    962
  ×  45
  ....
  ....0
  ____
  ....
```

Work out 962×5

Work out 962×4

(b) $442 \div 13$

$$13\overline{)442}\ \ 3...$$

(2 marks) (2 marks)

5 Dylan packs tomato tins into boxes. Each box holds 36 tomato tins. How many boxes will he need to pack

(a) 180 tins

(b) 324 tins?

.... (1 mark) .... (1 mark)

6 Sam bought five boxes of chocolates. Each box contained 25 chocolates. Sam ate 30 chocolates himself. He then shared the remaining chocolates between himself and his four friends.

(a) How many chocolates did Sam buy?

(b) How many chocolates did each of Sam's friends receive?

.... (1 mark) .... (2 marks)

NUMBER — Had a go ☐ Nearly there ☐ Nailed it! ☐

Decimals and place value

Target grade 1

1 (a) Write down the value of the 7 in 9.74

.... (1 mark)

> Remember the first number after the decimal point is a tenth and then a hundredth and so on.

(b) Write down the value of the 8 in 0.684

.... (1 mark)

(c) Write down the value of the 4 in 0.704

.... (1 mark)

Target grade 1

2 Write the following numbers in order, smallest first:

3.2   6.4   6.2   12.8   1.4

.... (1 mark)

Target grade 1 — Guided

3 Write the following numbers in order, smallest first:

0.61   0.611   0.613   0.6   0.05

0.610   0.611   0.613   0.600   0.050̶

> Place zeros on these numbers so they all have the same number of decimal places.

0.050   ....   (1 mark)

Target grade 1

4 Write the following numbers in order, smallest first:

0.73   0.7   0.725   0.778   0.78

.... (1 mark)

Target grade 2 — Guided

5 Using the information that $5.7 \times 43 = 245.1$ write down the value of

(a) $57 \times 43 =$

245.1 × .... = ...

> 5.7 has been multiplied by 10 and 43 is unchanged. 245.1 needs to be multiplied by 10.

(1 mark)

(b) $5.7 \times 4.3 =$

245.1 ÷ .... = ...

> 5.7 is unchanged and 43 has been divided by 10. 245.1 needs to be divided by 10.

(1 mark)

(c) $245.1 \div 57 =$

43 ÷ .... = ...

> 245.1 is unchanged and 5.7 has been multiplied by 10. 43 needs to be divided by 10.

(1 mark)

Target grade 2

6 Sammy writes down the following in his exercise book

$$435.2 \div 13.6 = 320$$

He uses the information to say that $32 \times 136 = 4352$

Is he correct? Explain your answer.

> Write 'yes' or 'no' and give your reason. You can use working to explain your answer if it is easier than writing it as a sentence.

.... (1 mark)

Had a go ☐ Nearly there ☐ Nailed it! ☐   **NUMBER**

Operations on decimals

1 Work out

(a) $4.23 + 10.4$

```
   4.23
+ 10.40
 ──────
   ....3
```

Make sure all the decimal points are lined up and then write zeros in the spaces so that all the numbers have the same number of decimal places.

(b) $84.7 - 9.34$

```
  84.70
-  9.34
 ──────
  .....
```

(2 marks)   **(2 marks)**

(c) 7.32×16

First work out 732×16. In total there are 2 decimal places in the calculation, so put 2 decimal places in your answer.

(d) 0.47×0.07

.... **(2 marks)**   .... **(2 marks)**

(e) $83.4 \div 6$

$6\overline{)83.4}$ .... **(2 marks)**

(f) $81.9 \div 1.3$

.... **(2 marks)**

2 A coach ticket to the zoo costs £7.85. A teacher buys 36 of these tickets for his class. What is the total cost of the 36 tickets?

```
   785
 ×  36
 ─────
```

Total cost = £.....................

In total there are 2 decimal places in the calculation, so put 2 decimal places in your answer. Remember to write in the units.

(2 marks)

3 Charles repairs computers.
He charged a customer £123.20 to repair a computer.
It took him 8 hours to repair the computer.
How much did he charge for one hour?

£......................... **(2 marks)**

4 Kitty buys hot chocolate sachets.
There are 14 hot chocolate sachets in a small box.
A small box costs £3.49.
Kitty uses 3 hot chocolate sachets each day.
Work out the how much Kitty spends on hot chocolate sachets in a four-week period.

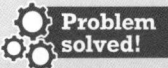 For a longer question like this, it's a good idea to plan your strategy. Calculate
1. number of days in a four-week period
2. number of sachets used in a four-week period
3. number of small boxes used in a four-week period
4. total cost of those boxes.

£......................... **(4 marks)**

7

NUMBER

Had a go ☐   Nearly there ☐   Nailed it! ☐

Squares, cubes and roots

1 Work out

(a) 4^2

.... **(1 mark)**

(b) 2^3

.... **(1 mark)**

(c) $\sqrt{81}$

.... **(1 mark)**

(d) $\sqrt{64}$

.... **(1 mark)**

(e) $\sqrt[3]{64}$

.... **(1 mark)**

(f) $\sqrt[3]{8}$

.... **(1 mark)**

(g) $\sqrt[3]{27}$

.... **(1 mark)**

(h) $\sqrt[3]{-64}$

.... **(1 mark)**

(i) $\sqrt[3]{-125}$

.... **(1 mark)**

2 Write down

(a) the square of 9

.... **(1 mark)**

(b) the cube of 5

.... **(1 mark)**

(c) the square root of 144

.... **(1 mark)**

(d) the cube root of 216

.... **(1 mark)**

3 Work out the value of $5^2 + 3^3$

> Square the 5 and cube the 3 before you add.

 $(5 \times 5) + (3 \times 3 \times 3) =$ .... + ... = ... **(1 mark)**

4 2, 8, 11, 15, 21, 26, 36, 49

Write down a number from the list that

(a) is a square number

(b) is a cube number

(c) has a square root of 7.

.... **(1 mark)**   .... **(1 mark)**   .... **(1 mark)**

5 Tom carried out an investigation and concluded that

'6 is a cube number since $2^3 = 6$'

Is he correct?

> You can explain your answer by writing a sentence with your reason, or by showing some neat working.

No, because $2 \times 2 \times 2 =$ .... **(1 mark)**

6 If you add three square numbers then you always get an even number.

Is this statement correct? Explain your answer.

> **Problem solved!** Try to find an example of three square numbers that add up to an **odd** number to show the statement is incorrect.

.... **(1 mark)**

Had a go ☐ **Nearly there** ☐ **Nailed it!** ☐

NUMBER

Indices

Target grade 3

1 Write as a single power of 4

Guided

(a) $4 \times 4 = 4^{\cdots}$ **(1 mark)** (b) $4 \times 4 \times 4 \times 4 \times 4 = \ldots\ldots\ldots$ **(1 mark)**

Target grade 4

2 Simplify and leave your answers in index form.

Guided

(a) $5^3 \times 5^6$ — Add the powers.

$5^3 \times 5^6 = 5^{3+6} = 5^{\cdots}$ **(1 mark)**

(b) $5^9 \div 5^6$ — Subtract the powers.

$5^9 \div 5^6 = 5^{9-6} = 5^{\cdots}$ **(1 mark)**

(c) $\dfrac{5^{12}}{5 \times 5^7}$ — First work out the power of 5 in the denominator.

(d) $(5^3)^4$ — Multiply the powers.

$\ldots\ldots\ldots\ldots$ **(2 marks)** $\ldots\ldots\ldots\ldots$ **(1 mark)**

Target grade 5

3 Write as a single power of 9

Guided

(a) $\dfrac{1}{9} = 9^{\cdots}$ **(1 mark)** (b) $\dfrac{1}{9 \times 9 \times 9 \times 9} = \ldots\ldots\ldots\ldots$ **(1 mark)**

Target grade 5

4 Simplify and leave your answers in index form.

(a) $\dfrac{3^2 \times 3^6}{3^5}$ (b) $\dfrac{3^{12}}{3^6 \times 3^4}$ (c) $\dfrac{3^7 \times 3^6}{3 \times 3^4}$ (d) $\dfrac{3^8 \times 3^{-6}}{3 \times 3^{-5}}$

$\ldots$ **(2 marks)** $\ldots$ **(2 marks)** $\ldots$ **(2 marks)** $\ldots$ **(2 marks)**

Target grade 5

5 Complete the following:

Anything to the power zero equals ONE.

(a) $7^0 = \ldots\ldots$ **(1 mark)** (b) $7^{-1} = \dfrac{1}{\ldots\ldots}$ **(1 mark)**

(c) $7^{-2} = \dfrac{1}{7^2} = \dfrac{1}{\ldots\ldots}$ **(1 mark)** (d) $4^{-3} = \ldots\ldots$ **(1 mark)**

(e) $\left(\dfrac{3}{4}\right)^3 = \dfrac{3^3}{4^3} = \dfrac{\ldots\ldots}{\ldots\ldots}$ **(1 mark)** (f) $\left(\dfrac{4}{5}\right)^{-2} = \left(\dfrac{5}{4}\right)^2 = \dfrac{\ldots\ldots}{\ldots\ldots}$ **(1 mark)**

Turn the fraction upside down, then change the negative power to a positive power.

Target grade 5

6 $7^4 \times 7^x = \dfrac{7^9 \times 7^6}{7^3}$

Find the value of x.

$x = \ldots\ldots\ldots\ldots$ **(2 marks)**

9

NUMBER

Had a go ☐ Nearly there ☐ Nailed it! ☐

Estimation

1 Work out an estimate for the value of

(a) $188 \times 69 \approx 200 \times 70 =$ ....

Round both values to 1 significant figure. **(1 mark)**

(b) $28.9 \div 4.85 \approx$ .... $\div$ .... = ... **(1 mark)**

(c) $(51.2)^3 \approx ($.....................$)^3 =$ .... **(1 mark)**

2 Work out an estimate for the value of $\dfrac{4826}{4.1 \times 9.72}$

$\approx \dfrac{5000}{4 \times \text{.........}} = \dfrac{\text{.........}}{\text{.........}} =$ ....

1. Round all values to 1 significant figure.
2. Multiply the numbers in the denominator.
3. Cancel if possible, then divide.

(2 marks)

3 Work out an estimate for the value of $\dfrac{8.92 \times 408}{0.506}$

.... **(2 marks)**

4 Work out an estimate for the value of $\dfrac{716 \times 5.13}{0.191}$

$\dfrac{700 \times 5}{0.2} = \dfrac{3500}{0.2} = \dfrac{\text{.........}}{2} =$ ....

If you need to divide by a decimal you can multiply top and bottom by 10 or 100 to simplify the calculation. **(2 marks)**

5 Work out an estimate for the value of $\dfrac{29 \times 4.90}{0.204}$

.... **(2 marks)**

6 Work out an estimate for the value of $\dfrac{5.89 \times 291}{0.051}$

.... **(2 marks)**

7 The radius of a sphere is 6.2 cm.

Surface area of a sphere = $4\pi r^2$

(a) Work out an estimate for the surface area of the sphere.

Examiners' report: If you are multiplying (or squaring) a number, then rounding down will produce an underestimate, and rounding up will produce an overestimate.

Round π and r to 1 significant figure.

.... cm² **(2 marks)**

(b) Without further calculation, explain whether your method gives you an overestimate or an underestimate for the surface area of the sphere.

.... **(1 mark)**

Had a go ☐ Nearly there ☐ Nailed it! ☐   NUMBER

Factors, multiples and primes

Target grade 1

1 (a) Write down all the factors of 36.

> Guided

   1 × 36, 2 × ...., ... × ...., ... × ...., ... × ....   **(2 marks)**

   (b) Write down the first ten multiples of 7.

   7   14   ....   **(1 mark)**

Target grade 1

2 Use a word from the box to complete these sentences correctly.

   | multiple | factor |
   | square root | cube |

   (a) 12 is a .... of 132.   **(1 mark)**

   (b) 132 is a .... of 12.   **(1 mark)**

Target grade 1

3 The table shows some numbers.

41	42	43	44	45	46	47	48	49

   Three of the numbers are prime numbers.
   Put a tick (✓) underneath each of these three numbers.   **(1 mark)**

Target grade 1

4 From this list of numbers write down

   2   8   6   12   21   25   33   49

   (a) a factor of 30 .... **(1 mark)**   (b) a multiple of 7 .... **(1 mark)**

   (c) two factors of 24 that have a product of 48 ....   **(2 marks)**

Target grade 1

5 Write down three factors of 28 which have a sum between 20 and 25.

   > Start by listing the factors of 28.

   .... **(2 marks)**

Target grade 4

6 Express the following numbers as products of their prime factors.
   Give your answers in index form.

> Guided

   (a) 54

   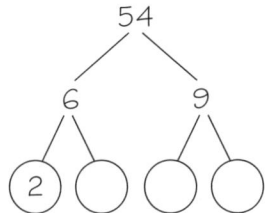

   54 = 2 × .... × .... × .... = 2 × ....^....   **(3 marks)**

   > The prime factors are always circled.

   (b) 96   (c) 126   (d) 252

   .... **(3 marks)**   .... **(3 marks)**   .... **(3 marks)**

NUMBER

Had a go ☐ Nearly there ☐ Nailed it! ☐

HCF and LCM

1 (a) Find the highest common factor (HCF) of 72 and 84.

> 1. List the factors of 72.
> 2. List the factors of 84.
> 3. Circle all the common factors.
> 4. Choose the highest common factor.

 Guided

1 × 72, 2 × ...., ... × ...., ... × ...., ... × ...., ... × ....

1 × 84, 2 × ...., ... × ...., ... × ...., ... × ...., ... × ....

.... **(3 marks)**

(b) Find the lowest common multiple (LCM) of 12 and 15.

.... **(2 marks)**

2 (a) Write the following numbers as products of their prime factors.

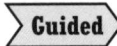 Guided

(i) 90

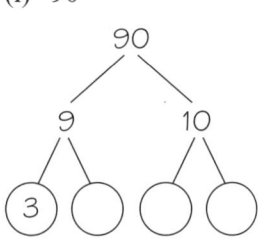

> **Examiners' report** You need to include multiplication signs in your answer.

90 = 3 × .... × .... × .... **(2 marks)**

(ii) 210

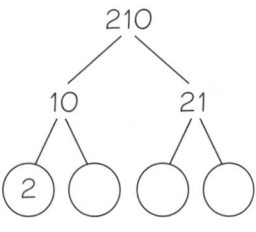

210 = 2 × .... **(2 marks)**

(b) Find the highest common factor (HCF) of 90 and 210.

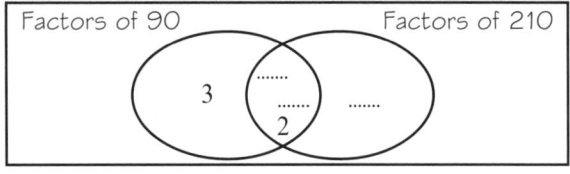

> Draw a Venn diagram showing the prime factors of each number. Multiply the prime factors in the **intersection** of the Venn diagram.

HCF = 2 × .... × .... = ... **(1 mark)**

(c) Find the lowest common multiple (LCM) of 90 and 210.

> Multiply **all** the prime factors in the Venn diagram.

LCM = 3 × .... × .... × .... × .... = ... **(1 mark)**

3 (a) Find the highest common factor (HCF) of 36 and 48.

.... **(1 mark)**

(b) Find the lowest common multiple (LCM) of 36 and 48.

.... **(1 mark)**

Had a go ☐ Nearly there ☐ Nailed it! ☐   **NUMBER**

Fractions

Target grade 1

1 Shade $\frac{3}{7}$ of this shape.

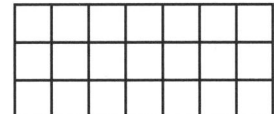

(1 mark)

> **Examiners' report** Learn your times tables up to 12 × 12. It makes fraction questions much easier!

Target grade 1

2 Write the following fractions in their simplest form.

> Guided

(a) $\frac{30}{60} = \text{................}$ **(1 mark)**

(b) $\frac{12}{18} = \text{................}$ **(1 mark)**

(c) $\frac{35 \div 5}{120 \div 5} = \frac{}{}$ *What number will go into 35 and 120?* **(1 mark)**

(d) $\frac{24}{84} = \text{................}$ **(1 mark)**

Target grade 1

3 Write down the fraction of these shapes that are shaded. Write your fraction in its simplest form.

(a)    .... **(2 marks)**

(b) 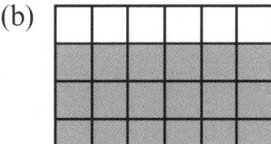   .... **(2 marks)**

Target grade 1

4 Work out

> Guided

(a) $\frac{3}{4}$ of £60

£60 ÷ 4 = ...

.... × 3 = £.................. **(2 marks)**

(b) $\frac{4}{5}$ of £80

£.................. **(2 marks)**

(c) $\frac{7}{8}$ of £160

£.................. **(2 marks)**

(d) $\frac{5}{7}$ of £210

£.................. **(2 marks)**

Target grade 2

5 Sandeep bought 30 punnets of strawberries for £20.

$\frac{1}{6}$ of the punnets were rotten so he threw them away.

He sold the remaining punnets for £1.50 each.
Work out Sandeep's profit.

> **Problem solved!** **Profit** means the total amount of money Sandeep made, minus the £20 that he spent.

£.......................... **(4 marks)**

Target grade 2

6 Tom bought 20 boxes of flowers. Each box cost him £6.
Each box contains 15 flowers.

He sells $\frac{3}{5}$ of the total number of flowers for 70p each.

He then sells the remaining at 50p each.
Work out the total profit Tom makes.

£........................ **(5 marks)**

13

NUMBER   Had a go ☐   Nearly there ☐   Nailed it! ☐

Operations on fractions

Target grade 2

Guided

1. Work out   [Write both fractions as equivalent fractions with the same denominator.]

   (a) $\dfrac{1}{3} + \dfrac{2}{5}$

   $= \dfrac{5}{15} + \dfrac{\ldots}{15} = \dfrac{\ldots}{15}$ **(2 marks)**

   (b) $\dfrac{4}{5} - \dfrac{1}{4}$

   $= \dfrac{\ldots}{20} - \dfrac{\ldots}{20} = \dfrac{\ldots}{20}$ **(2 marks)**

   (c) $\dfrac{6}{7} + \dfrac{3}{8}$

   .... **(2 marks)**

   (d) $\dfrac{5}{9} - \dfrac{4}{7}$

   .... **(2 marks)**

Target grade 2

2. Work out

   (a) $\dfrac{1}{2} \times \dfrac{1}{3}$

   .... **(1 mark)**

   (b) $\dfrac{5}{11} \times \dfrac{3}{4}$

   .... **(1 mark)**

   (c) $\dfrac{4}{5} \div \dfrac{3}{10}$ [Turn the second fraction upside down and change ÷ into ×]

   .... **(2 marks)**

   (d) $\dfrac{2}{3} \div \dfrac{4}{9}$

   .... **(2 marks)**

Target grade 2

Guided

3. A man wins some money and decides to give it to his three children.
   Andrew receives $\dfrac{2}{5}$ of the money, Ben receives $\dfrac{1}{3}$ of the money and Carla receives the rest.
   Work out the fraction that Carla receives.

   $\dfrac{2}{5} + \dfrac{1}{3} = \dfrac{\ldots}{15} + \dfrac{\ldots}{15} = \dfrac{\ldots}{15}$

   [Write 1 as a fraction with the same numerator and denominator. $1 = \dfrac{15}{15}$]

   $1 - \dfrac{\ldots}{15} = \dfrac{\ldots}{15} =$ **(3 marks)**

Target grade 2

4. A garage has a supply of 210 litres of oil.
   Amy uses $\dfrac{4}{7}$ of the supply and Brad uses $\dfrac{1}{5}$ of the supply.

   (a) What fraction of the supply is left?

   .... **(3 marks)**

   (b) How much oil is left?

   .... litres **(2 marks)**

14

Had a go ☐   Nearly there ☐   Nailed it! ☐   **NUMBER**

Mixed numbers

Tricky Topic

1 Work out

(a) $3\frac{4}{5} + 2\frac{3}{4}$

> You need to write mixed numbers as improper fractions before you do any calculations.

$= \frac{19}{5} + \frac{.....}{4} = \frac{.....}{20} + \frac{.....}{20} = \frac{.....}{20} = ...$

> Write your final answer as a mixed number in its simplest form.

(3 marks)

(b) $4\frac{2}{5} - 2\frac{3}{10}$

$= \frac{.....}{5} - \frac{.....}{10} = \frac{.....}{10} - \frac{.....}{10} = \frac{.....}{10} = ...$ **(3 marks)**

2 Work out

(a) $1\frac{2}{3} \times 2\frac{3}{10}$

$= \frac{.....}{3} \times \frac{.....}{10} = \frac{.....}{......} = ...$ **(3 marks)**

(b) $4\frac{2}{3} \div 1\frac{2}{5}$

$= \frac{.....}{3} \div \frac{.....}{5} = \frac{.....}{3} \times \frac{.....}{......} = \frac{.....}{......} = ...$

> Don't forget to replace the ÷ with × sign and then flip the fraction over.

(3 marks)

3 Work out

(a) $3\frac{1}{2} \times 2\frac{4}{7}$ (b) $5\frac{1}{3} \div 1\frac{4}{9}$

.... **(3 marks)**   .... **(3 marks)**

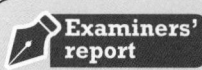

4 It takes $4\frac{2}{3}$ hours to paint a room, and $1\frac{1}{4}$ hours for all the paint to dry.

How long does it take altogether?

> **Examiners' report** You can check your answer makes sense by estimating. $4\frac{2}{3} + 1\frac{1}{4}$ is close to $5 + 1 = 6$

.... hours **(3 marks)**

5 James has $1\frac{1}{6}$ litres of milk left in the fridge and needs $2\frac{2}{9}$ litres for a recipe.

How much more milk does he need?

.... litres **(3 marks)**

6 A tape is $14\frac{2}{3}$ m long. How many pieces of tape each of $1\frac{2}{9}$ m can be cut from the length of tape?

.... **(3 marks)**

15

NUMBER

Had a go ☐ Nearly there ☐ Nailed it! ☐

Calculator and number skills

1 Work out

(a) $11 + 8 \div 2$

$11 + \ldots\ldots = \ldots\ldots$ **(1 mark)**

(b) $2 + 9 \times 10 + 3$

$\ldots\ldots\ldots\ldots$ **(1 mark)**

(c) $8 + (3 \times 20) \div 6$

$\ldots\ldots\ldots\ldots$ **(1 mark)**

(d) $(14 - 5)^2$

$\ldots\ldots\ldots\ldots$ **(1 mark)**

2 Work out

> You must use BIDMAS.

(a) $\dfrac{27 + 3 \times 3}{3 \times 2}$

$\ldots\ldots\ldots\ldots$ **(1 mark)**

(b) $\dfrac{13 - 12 \div 4}{4 + 3 \times 2}$

$\ldots\ldots\ldots\ldots$ **(1 mark)**

(c) $\dfrac{12 + 3 \times 6}{4 + 3 \div 3}$

$\ldots\ldots\ldots\ldots$ **(1 mark)**

3 Find the value of $\dfrac{4.5 + 3.75}{3.2^2 - 5.53}$

Write down all the figures on your calculator display.

$\dfrac{8.25}{\ldots\ldots\ldots}= \ldots\ldots$ **(2 marks)**

4 (a) Find the value of $\sqrt{30.25} + 1.75^2$

> Enter the numbers into the calculator. You might need to press the 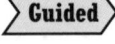 button to get your answer as a decimal number.

$\ldots\ldots\ldots\ldots\ldots\ldots\ldots\ldots\ldots\ldots\ldots\ldots\ldots$ **(2 marks)**

(b) Write your answer to part (a) correct to one significant figure.

$\ldots\ldots\ldots\ldots$ **(1 mark)**

5 (a) Find the value of $\dfrac{\sqrt{18.3 + 3.6^2}}{2.8 \times 1.6}$

Write down all the figures on your calculator display.

> 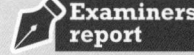 Work out the numerator and denominator separately, and write them down. Then work out the answer, and write down **all** the figures on your calculator.

$\ldots\ldots\ldots\ldots$ **(2 marks)**

(b) Write your answer to part (a) correct to 3 significant figures.

$\ldots\ldots\ldots\ldots$ **(1 mark)**

6 (a) Find the value of $\dfrac{32.5 \times \sqrt[3]{16.3}}{9.5 \times 3.1}$

Write down all the figures on your calculator display.

$\ldots\ldots\ldots\ldots$ **(2 marks)**

(b) Write your answer to part (a) correct to 2 significant figures.

$\ldots\ldots\ldots\ldots$ **(1 mark)**

Had a go ☐ Nearly there ☐ Nailed it! ☐   **NUMBER**

Standard form 1

1 (a) Write 45 000 in standard form.

45 000 = 4.5 × 10......

Count decimal places from the right. How many jumps do you need to make to get 4.5?

(1 mark)

(b) Write 3.4×10^{-5} as an ordinary number.

3.4×10^{-5} = 0.0.........

The power of 10 is negative so the number is less than 1.

(1 mark)

(c) Write 28×10^6 in standard form.

.... (1 mark)

2 Write in standard form

(a) 567 000

(b) 0.000 056 7

(c) 567×10^8

.... (1 mark)   .... (1 mark)   .... (1 mark)

3 (a) Write 6 740 000 in standard form.

.... (1 mark)

$n = 6\,740\,000$ and $m = 5.42 \times 10^5$

Work out, giving your answers in standard form correct to 2 significant figures,

Use the ×10ˣ button to enter standard form numbers on your calculator.

(b) $n + m$

6 740 000 + ...

= .... (2 marks)

(c) $n - m$

6 740 000 − ....

= .... (2 marks)

4 In 2014 the population of the United Kingdom was 6.5×10^7
In 2014 the population of Russia was 1.4×10^8

(a) Work out the combined population of the United Kingdom and Russia. Give your answer in standard form.

.... (2 marks)

(b) Work out the difference between the population of the United Kingdom and the population of Russia. Give your answer in standard form.

.... (2 marks)

5 In 2011, NASA launched the spacecraft Curiosity to land on the planet Mars. The distance from Earth to Mars is 5.63×10^8 km. The time it took to reach Mars was 6050 hours.
Work out the average speed, in km/h, of the spacecraft Curiosity.
Give your answer in standard form correct to 2 significant figures.

$\text{speed} = \dfrac{\text{distance}}{\text{time}} = $ .... (2 marks)

Standard form 2

6 Work out, giving your answers in standard form,

(a) $(3 \times 10^6) \times (6 \times 10^{-3})$

$= (3 \times \text{............}) \times (10^6 \times 10^{\text{......}}) = \text{............} \times 10^{\text{......}} = \text{............} \times 10^{\text{......}}$ **(2 marks)**

(b) $(8 \times 10^6) \div (4 \times 10^{-14})$

$= (8 \div \text{............}) \times (10^6 \div 10^{\text{......}}) = \text{............} \times 10^{\text{......}}$ **(2 marks)**

7 Work out, giving your answers in standard form,

(a) $5.1 \times 10^3 + 6.5 \times 10^4$

    5 1 0 0
+ 65 000

.... **(2 marks)**

(b) $7.6 \times 10^5 - 8 \times 10^3$

    760 000
−   8 000

.... **(2 marks)**

8 A and B are standard form numbers.
$A = 5.6 \times 10^9 \qquad B = 8 \times 10^{-2}$
Calculate, giving your answers in standard form,

(a) $2A$ (b) $A \times B$ (c) $A \div B$

.... **(2 marks)**   .... **(2 marks)**   .... **(2 marks)**

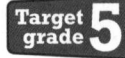

9 It takes light 8 minutes to travel from the Sun to the Earth.
The speed of light is 3×10^8 m/s.
Work out the distance, in km, from the Sun to the Earth.
Give your answer in standard form.

$$\text{speed} = \frac{\text{distance}}{\text{time}}$$

.... **(3 marks)**

10 The distance from the Sun to the planet Neptune is approximately 4.5×10^9 km.
The speed of light is 3×10^8 m/s.
Work out how long, in seconds, it takes light to travel from the Sun to the planet Neptune.

Problem solved! Convert the distance into metres, then use $\text{time} = \dfrac{\text{distance}}{\text{speed}}$

.... **(3 marks)**

Had a go ☐ Nearly there ☐ Nailed it! ☐   **NUMBER**

Counting strategies

Target grade 3

Guided

1 Ajay writes down one letter from the word ART and then he writes down one number from 1, 2 and 3.

| A | R | T | | 1 | 2 | 3 |

Do not write down repeats such as (1, A) which is the same as (A, 1).

List all the possible combinations Ajay could write down. One has been done for you.

(A,1) (A,..........) (A,..........) (R,..........) (R,..........) (R,..........)
(T,..........) (T,..........) (T,..........)

(2 marks)

Target grade 3

2 Brett goes to a restaurant.
He can choose from three types of curry and from three types of naan.
Brett is going to choose one curry and one naan.
List all the possible combinations Brett can choose.

Curry	Naan
Chicken (C)	Plain
Lamb	Garlic
Vegetable	Butter

Label chicken as C, and so on.

.... **(2 marks)**

Target grade 4

3 Emily has four tiles.
One tile is marked W, one tile is marked X, one tile is marked Y and one tile is marked Z.
Emily chooses two of these tiles.
Write down all the possible combinations she can get.

| W | X | Y | Z |

.... **(2 marks)**

Target grade 4

4 Kate has three cards. Each card has a different digit on it.
Kate wants to make a three-digit number.
Each number is made with all three cards.
How many different numbers can Kate make?

| 3 | 6 | 9 |

.... **(2 marks)**

Target grade 4

Guided

5 There are four players in a competition, Asha, Bev, Chloe and Dan.
Each player must play each other once.
How many games will be played in total?

Problem solved! Label Asha, Bev, Chloe and Dan as A, B, C and D respectively. Remember (A, B) is the same as (B, A).

(A,..........) (A,..........) (A,..........) (B,..........) ....

.... **(2 marks)**

Target grade 4

6 A school chess league contains the following five teams:
Hanover, Norman, Stuart, Tudor and Windsor.
Each team must play each other **twice**.
How many games will be played in total?

.... **(2 marks)**

Problem-solving practice 1

NUMBER — Had a go ☐ Nearly there ☐ Nailed it! ☐

1 Find four different prime numbers you can add together to get a number greater than 30 and less than 40.

.... **(2 marks)**

2 Crisps cost 35p per packet. A bottle of lemonade costs £1.25. Nigel buys five packets of crisps and one bottle of lemonade. He pays with a £10 note.
Work out how much change he should get.

£................................ **(3 marks)**

3 Here is part of a menu in Harry's café.

Abbie buys some cups of coffee. She has £15.
Work out the greatest number of cups of coffee she can buy.

| Cup of tea | £1.30 |
| Cup of coffee | £1.40 |

.... **(2 marks)**

4 A shop sells packets of sweets. There are 36 packets of sweets in each box.
In November, the shop sold all the packets of sweets in 120 boxes.
In December, the shop sold all the packets of sweets in 230 boxes.

(a) Work out the total number of packets of sweets the shop sold.

.... **(2 marks)**

(b) Vans deliver the boxes to the shop. A van can carry 72 boxes.
Sandra wants 452 boxes. Sandra works out she needs 6 vans to deliver the boxes. Is she correct? You must show all your working.

.... **(2 marks)**

5 Which of these fractions is the larger: $\frac{2}{3}$ or $\frac{3}{5}$?
You must show clearly how you got your answer.

(3 marks)

6 A machine makes 48 bolts every hour. The machine makes bolts for $7\frac{1}{2}$ hours each day, on 5 days of the week. The bolts are packed into boxes. Each box holds 30 bolts. How many boxes are needed for all the bolts made each week?

.... boxes **(4 marks)**

Had a go ☐   Nearly there ☐   Nailed it! ☐   **NUMBER**

Problem-solving practice 2

7 There are 180 counters in a bag. The counters are brown or white or blue.
$\frac{3}{5}$ of the counters are brown. $\frac{1}{4}$ of the counters are white.
Work out the number of blue counters.

.... **(4 marks)**

8 Tammy buys 3 compost bags. Each compost bag weighs 24 kg.
She can fill a small pot by using $\frac{4}{9}$ of a compost bag.
How many pots can she fill?

.... **(3 marks)**

9 A plumber has two lengths of copper pipe.
One pipe is 54 cm and the other pipe is 72 cm.
He wants to cut them and make smaller pipes to use them in boilers.
He wants the smaller pipes to be the same length with no copper left over.
What is the greatest length, in cm, that he can make the smaller pipes?

.... cm **(3 marks)**

10 Buses leave Wolverhampton for Penn every 12 minutes and for Wombourne every 15 minutes.
A bus to Penn and a bus to Wombourne leave Wolverhampton at 8 am.
At what time will a bus to Penn and a bus to Wombourne next leave Wolverhampton at the same time?

.... **(3 marks)**

11 An atomic particle has a lifetime of 4.86×10^{-5} seconds.
It travels at a speed of 6.2×10^{4} m/s. Show that the distance travelled by the atomic particle is approximately 3 m.

(2 marks)

12 Find the value of x.

$$3^x \times 3^{4x} = \frac{3^7 \times 3^9}{3^2}$$

$x =$ .... **(2 marks)**

21

ALGEBRA

Had a go ☐   Nearly there ☐   Nailed it! ☐

Collecting like terms

1   formula   equation   expression   term

Choose a word from the list above to best describe each of the following:

(a) $3x + y$ (b) $v = u + at$ (c) $3x + 4 = 6$

.... **(1 mark)**   .... **(1 mark)**   .... **(1 mark)**

2 Simplify

(a) $x + x + x + x + x$

= .... x **(1 mark)**

(b) $3xy + 5xy - 2xy$

= .... xy **(1 mark)**

3 Simplify

(a) $x + y + x + y + x + x$

= .... x + ... y **(1 mark)**

(b) $8ab - 3ab$

= .... ab **(1 mark)**

(c) $5t + 6v - 4t + 5v$

= $5t - 4t + 6v + 5v$ = ... t + ... v **(1 mark)**

(d) $6c + 2d - 3c - 4d$

= $6c - 3c + 2d - 4d$ = ... c - .... d **(1 mark)**

4 Simplify

(a) $5x - 3x$

.... **(1 mark)**

(b) $4t^2 - t^2$

> **Examiners' report** This means 'four lots of t^2 minus one lot of t^2'.

.... **(1 mark)**

(c) $4a + 5b - a + 3b + 7$

.... **(2 marks)**

(d) $6x - 3y - 5x - 4y$

.... **(2 marks)**

5 Simplify

(a) $5n - 2n$

.... **(1 mark)**

(b) $6p + 8q + 2q - 9q$

.... **(1 mark)**

(c) $3m - 4n + 6n + 2n$

.... **(2 marks)**

(d) $5 + 3a + 7b - a - b$

.... **(2 marks)**

6 Simplify

(a) $y^2 + y^2$

.... **(1 mark)**

(b) $5x^2 + 3x - 2x^2 - x$

.... **(2 marks)**

Had a go ☐   Nearly there ☐   Nailed it! ☐   **ALGEBRA**

Simplifying expressions

Target grade 2
Guided

1 Simplify

(a) $y \times y$

$= y^{......}$ **(1 mark)**

(b) $3m \times t$

$= 3 \times m \times t = $ .... **(1 mark)**

Target grade 2

2 Simplify

(a) $w \times w \times w \times w$

.... **(1 mark)**

(b) $4 \times 7 \times d$

.... **(1 mark)**

(c) $5 \times 6k$

.... **(1 mark)**

(d) $5j \times 8k$

.... **(1 mark)**

Target grade 2
Guided

3 Simplify

(a) $5x \times 3x$

$= 5 \times 3 \times x \times x = $ .... **(1 mark)**

(b) $2e \times 3f$

$= 2 \times 3 \times e \times f = $ .... **(1 mark)**

(c) $8a \div 2$ [Work out $8 \div 2$]

$= \dfrac{8a}{2} = $ .... **(1 mark)**

(d) $24ab \div 3a$ [Divide the number parts and then the letter parts.]

$= \dfrac{24ab}{3a} = $ .... **(1 mark)**

Target grade 2

4 Simplify

(a) $7g \times 5h$

.... **(1 mark)**

(b) $2t \times 2t \times 2t$

.... **(1 mark)**

(c) $36xy \div 12y$

.... **(1 mark)**

(d) $28xyz \div 7xz$

.... **(1 mark)**

Target grade 2

5 Simplify

(a) $2a \times 3b \times 4c$

.... **(1 mark)**

(b) $48mnp \div 6mn$

.... **(1 mark)**

Target grade 2

6 The table shows some expressions.

$4x \div 2y$	$2(x \times y)$	$2x \times 2y$	$4xy \div 2$	$x \times y$

Two of the expressions always have the same value as $2xy$
Tick (✓) the boxes underneath the two expressions.   **(2 marks)**

23

ALGEBRA

Had a go ☐ Nearly there ☐ Nailed it! ☐

Algebraic indices

 **1** Simplify these expressions and leave your answers in index form:

(a) $a^3 \times a^6$ [Add the powers.]   (b) $a^9 \div a^6$ [Subtract the powers.]

.... **(1 mark)**   .... **(1 mark)**

(c) $\dfrac{a^{12}}{a \times a^7}$ [First work out the power of a in the denominator.]   (d) $(a^3)^4$ [Multiply the powers.]

.... **(2 marks)**   .... **(2 marks)**

 **2** Simplify these expressions and leave your answers in index form:

(a) $\dfrac{t^2 \times t^6}{t^5}$ (b) $\dfrac{t^{12}}{t^5 \times t^4}$

.... **(2 marks)**   .... **(2 marks)**

(c) $\dfrac{t^7 \times t^6}{t \times t^4}$ (d) $\dfrac{t^8 \times t^{-6}}{t \times t^{-5}}$

.... **(2 marks)**   .... **(2 marks)**

 **3** Simplify

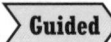

(a) $(x^3)^4$ (b) $(4x^2)^3$

.... **(1 mark)** $= 4x^2 \times 4x^2 \times 4x^2 = $ .... **(2 marks)**

(c) $(2x^3)^3$ (d) $3x^2 \times 4x^5$

.... **(2 marks)** $= 3 \times$ .... $\times x^{....+....} = $ .... **(2 marks)**

(e) $3x^2y \times 4x^5y^4$ (f) $18x^3y^5 \div 6xy^2$

.... **(1 mark)**   .... **(2 marks)**

 **4** Find the value of x.

(a) $p^4 \times p^x = p^{12}$ (b) $p^{12} \div p^x = p^7$ (c) $(p^3)^x = p^{15}$

$x = $ .... **(1 mark)** $x = $ .... **(1 mark)** $x = $ .... **(1 mark)**

 **5** Find the value of x.

$q^5 \times q^{2x} = \dfrac{q^{10} \times q^6}{q^8}$

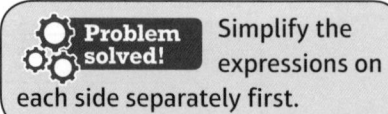

Simplify the expressions on each side separately first.

$x = $ .... **(2 marks)**

Had a go ☐   Nearly there ☐   Nailed it! ☐   **ALGEBRA**

Substitution

Target grade 1

1 The rule given in the box can be used to work out the distance a train travels.
A train has a speed of 80 km/h and travels for 3 hours.
Use the rule to work out the distance the train travels.

| distance = speed × time |

.... km   **(1 mark)**

Target grade 3

2 Work out the value of

| Use BIDMAS to find the correct value. |

Guided

(a) $5x + 3$ when $x = 4$

$= 5 \times \text{.......} + 3 = \text{.......}$ **(2 marks)**

(b) $2x - 3$ when $x = -4$

$= 2 \times \text{.......} - 3 = \text{.......}$ **(2 marks)**

Target grade 3

3 Work out the value of

(a) $4a + 3b$ when $a = 4$ and $b = 6$

.... **(2 marks)**

(b) $3a - 5b$ when $a = 3$ and $b = -2$

.... **(2 marks)**

Target grade 3

4 Work out the value of

| Use BIDMAS to find the correct value. |

Guided

(a) $3a + ax$ when $a = 3$ and $x = -5$

$= 3 \times \text{.........} + \text{.........} \times \text{.........} = \text{.............................}$

| + × − = − |

(2 marks)

(b) $4t^2$ when $t = -5$

$= 4 \times (\text{.........})^2 = 4 \times \text{.........} \times \text{.........} = \text{.............................}$

| − × − = + |

(2 marks)

(c) $3g^2 - 5g$ when $g = 3$

$= 3 \times (\text{.........})^2 - 5 \times \text{.........} = \text{.............................}$

(2 marks)

Target grade 3

5 Work out the value of

(a) $4(2x - 4y)$ when $x = 3$ and $y = -5$

Examiners' report: Use brackets when you substitute a negative number.

.... **(2 marks)**

(b) $3m + 5(p - n)$ when $m = 6$, $n = 2$ and $p = 3$

.... **(2 marks)**

(c) $9t - \frac{1}{2}at^2$ when $a = 2$ and $t = 4$

.... **(2 marks)**

Target grade 3

6 Abbie and Lisa are trying to work out the energy of a ball when it is dropped.
They use the formula in the box where $m = 2$ and $v = 3$
Abbie works out the value of E to be 9
and Lisa works out the value of E to be 18.
Who is correct? Give a reason for your answer.

| $E = \frac{1}{2}mv^2$ |

You will need to show some working to justify your answer.

.... **(2 marks)**

25

ALGEBRA Had a go ☐ Nearly there ☐ Nailed it! ☐

Formulae

Target grade 1

Guided

1. The time in minutes needed to cook a joint of meat is given by the formula in the box.
Work out the time needed to cook a 5 kg joint of meat.

   Time = Weight in kg × 4 + 30

   Substitute the value for the weight of the joint into the formula.

   Use BIDMAS to find the correct value.

   Time = ... × 4 + 30 = ... **(2 marks)**

Target grade 1

2. Andy the carpenter charges £25 for each hour he works at a job plus £50 callout charge.
The amount Andy charges, in pounds, can be worked out using this formula.
Andy works for five hours at a job.
Work out how much Andy charges.

   Charge = Number of hours worked × 25 + 50

   £.......................... **(2 marks)**

Target grade 3

Guided

3. The height of a growing tree is given by the formula $h = 3t + 12$
Work out the value of h when $t = 6$

   Substitute the value of t into the formula.

   Use BIDMAS to find the correct value.

   h = 3 × .... + ... = ... = ... **(2 marks)**

Target grade 3

4. A formula involving force, mass and acceleration is $F = ma$
Work out the value of F when $m = 12$ and $a = 3$

   $F = $ .... **(2 marks)**

Target grade 3

5. This formula is used in physics to calculate impulse: $I = mu - mv$
Work out the value of I when $m = 6$, $u = 8$ and $v = 5$

   $I = $ .... **(2 marks)**

Target grade 3

Guided

6. A formula to work out the velocity of a ball is $v = u + at$
Work out the value of v when $u = -20$, $a = 9$ and $t = 8$

   Substitute the values of u, a and t into the formula.

   Use BIDMAS to find the correct value.

   v = ... + ... × .... = ... = ... **(2 marks)**

Target grade 3

7. You can use the formula in the box to convert degrees Celsius, C, into degrees Fahrenheit, F. Use the formula to convert $-20\,°C$ into °F.

   $F = 1.8C + 32$

   ....°F **(2 marks)**

Target grade 3

8. Given that $y = 3x^2 - 4x + 3$ show that when $x = -2$, the value of y is 23.

   The question says 'show that' so you need to write out all your working clearly.

   (2 marks)

26

Had a go ☐   Nearly there ☐   Nailed it! ☐   **ALGEBRA**

Writing formulae

Target grade 3 **Guided**

1. A can of lemonade costs *g* pence. A bag of sweets cost *h* pence.
Harry buys four cans of lemonade and five bags of sweets.
Write down a formula, in terms of *g* and *h*, for the total cost, *C* pence.

   $C = 4 \times$ .... $+ 5 \times$ .... $= 4$ .... $+ 5$ .... **(2 marks)**

Target grade 3

2. Aliya plays a game with black counters and white counters. The number of points for a black counter is 10 and the number of points for a white counter is 20.
Aliya has *m* black counters and *n* white counters.
Her total number of points is *S*.
Write down a formula for *S* in terms of *m* and *n*.

   [Your formula should start $S =$]

   .... **(2 marks)**

Target grade 3 **Guided**

3. Jayne the electrician charges £30 for each hour she works at a job and a callout charge of £50. Jayne works *n* hours at a job. She charges *P* pounds.
Write down a formula for *P* in terms of *n*.

   $P = n \times$ .... $+$ .... $=$ .... $+$ .... **(2 marks)**

Target grade 3

4. The cost, in £, of hiring a van can be worked out by using the following rule.

   [Add 5 to the number of days.
   Multiply the answer by 8.]

   The cost of hiring a van for *m* days is £*T*.
   Write down a formula for *T* in terms of *m*.

   .... **(2 marks)**

Target grade 3 **Guided**

5. The diagram shows the lengths of the sides of a quadrilateral. All the lengths are in cm.
Write down a formula, in terms of *x*, for the perimeter, *P* cm, of the quadrilateral.

   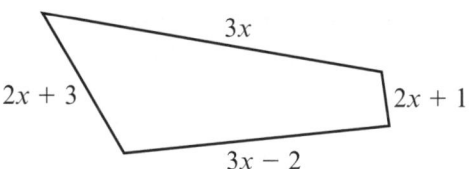

   Problem solved! The perimeter is the sum of the lengths of all the sides. Make sure you simplify as much as possible.

   $P = 2x + 3 +$ .... $+$ .... $+$ .... $=$ .... **(2 marks)**

Target grade 3

6. Alex is *n* years old. Brad is four years older than Alex.

   (a) Write down a formula, in terms of *n*, for Brad's age, *B*.

   .... **(1 mark)**

   The combined age of Alex, Brad and Carl is $5n + 4$

   (b) Is Carl three times older than Alex? Give a reason for your answer.

   .... **(2 marks)**

27

ALGEBRA

Had a go ☐ Nearly there ☐ Nailed it! ☐

Expanding brackets

1 Expand

(a) $3(x + 2)$ (b) $4(x + 5)$ (c) $5(x - 3)$

$= 3 \times x + 3 \times 2$

$= \ldots\ldots + \ldots\ldots$ **(2 marks)** $\ldots\ldots\ldots\ldots\ldots\ldots$ **(2 marks)** $\ldots\ldots\ldots\ldots\ldots\ldots$ **(2 marks)**

(d) $6(2x + 3)$ (e) $\sqrt{2}(x - \sqrt{2})$ (f) $7(3x - 8)$

$= 6 \times \ldots\ldots + 6 \times \ldots\ldots$

$= \ldots\ldots + \ldots\ldots$ **(2 marks)** $\ldots\ldots\ldots\ldots\ldots\ldots$ **(2 marks)** $\ldots\ldots\ldots\ldots\ldots\ldots$ **(2 marks)**

2 Expand *Multiply everything inside the brackets by **negative** 3. $-3 \times -3 = 9$ so you need to **add** 9 in part (a).*

(a) $-3(x - 3)$ (b) $-4(x + 3)$ (c) $-6(x - 5)$

$= -3 \times x - 3 \times -3$

$= \ldots\ldots + \ldots\ldots$ **(2 marks)** $\ldots\ldots\ldots\ldots\ldots\ldots$ **(2 marks)** $\ldots\ldots\ldots\ldots\ldots\ldots$ **(2 marks)**

(d) $-2(2x + 3)$ (e) $-2(4x - 1)$ (f) $-(2x - 4)$

$= -2 \times \ldots\ldots + -2 \times \ldots\ldots$

$= \ldots\ldots + \ldots\ldots$ **(2 marks)** $\ldots\ldots\ldots\ldots\ldots\ldots$ **(2 marks)** $\ldots\ldots\ldots\ldots\ldots\ldots$ **(2 marks)**

3 Expand

(a) $x(x + 1)$ (b) $x(x + 5)$ (c) $2x(x - 9)$

$= x \times x + x \times 1$

$= \ldots\ldots + \ldots\ldots$ **(2 marks)** $\ldots\ldots\ldots\ldots\ldots\ldots$ **(2 marks)** $\ldots\ldots\ldots\ldots\ldots\ldots$ **(2 marks)**

(d) $3x(2x - 3)$ (e) $-x(2x - 3)$ (f) $-3x(4x - 5)$

$= 3x \times \ldots\ldots + 3x \times \ldots\ldots$

$= \ldots\ldots + \ldots\ldots$ **(2 marks)** $\ldots\ldots\ldots\ldots\ldots\ldots$ **(2 marks)** $\ldots\ldots\ldots\ldots\ldots\ldots$ **(2 marks)**

4 Expand and simplify

(a) $4x + 3(x + 2)$ (b) $2(x + 1) + 3(x + 4)$

$= 4x + 3 \times x + 3 \times 2$ $= 2 \times x + 2 \times 1 + 3 \times x + 3 \times 4$

$= \ldots\ldots x + \ldots\ldots x + \ldots\ldots$ $= \ldots\ldots x + \ldots\ldots x + \ldots\ldots + \ldots\ldots$

$= \ldots\ldots x + \ldots\ldots$ **(3 marks)** $= \ldots\ldots x + \ldots\ldots$ **(3 marks)**

(c) $5(x - 3) + 4(2x + 1)$ (d) $4x(x - 3) + 2x(x - 4)$

$\ldots\ldots\ldots\ldots\ldots\ldots\ldots\ldots\ldots\ldots\ldots\ldots$ **(3 marks)** $\ldots\ldots\ldots\ldots\ldots\ldots\ldots\ldots\ldots\ldots\ldots\ldots$ **(3 marks)**

5 $3x(2x - 5) + 2x(7x - 2) = ax^2 + bx$ where a and b are whole numbers.
Work out the values of a and b.

$a = \ldots\ldots\ldots\ldots\ldots\ldots$

$b = \ldots\ldots\ldots\ldots\ldots\ldots$ **(3 marks)**

6 $5x(2 + 3x) - 4x(3x - 2) = px^2 + qx$ where p and q are whole numbers.
Work out the values of p and q.

$p = \ldots\ldots\ldots\ldots\ldots\ldots$

$q = \ldots\ldots\ldots\ldots\ldots\ldots$ **(3 marks)**

Had a go ☐ Nearly there ☐ Nailed it! ☐

ALGEBRA

Factorising

Target grade 3

Guided

1 Factorise

(a) $3x + 6$

$= 3(..... + ...)$ **(1 mark)**

(b) $6a + 18$

.... **(1 mark)**

(c) $2p - 6$

.... **(1 mark)**

(d) $5y - 15$

$= 5(..... - .....)$ **(1 mark)**

(e) $3t + 24$

.... **(1 mark)**

(f) $4g - 20$

.... **(1 mark)**

Target grade 3

Guided

2 Factorise

(a) $x^2 + 6x$

$= x(..... + ...)$ **(1 mark)**

(b) $x^2 - 4x$

.... **(1 mark)**

(c) $x^2 - 9x$

.... **(1 mark)**

(d) $x^2 - 12x$

$= x(..... - .....)$ **(2 marks)**

(e) $x^2 + 5x$

.... **(1 mark)**

(f) $x^2 - x$

.... **(1 mark)**

Target grade 4

Guided

3 Factorise fully

'Factorise fully' means that you need to take out the highest common factor.

(a) $3p^2 + 6p$

$= 3p(..... + ...)$ **(1 mark)**

(b) $8y^2 - 24y$

.... **(2 marks)**

(c) $9t^2 - 36t$

.... **(2 marks)**

Target grade 4

4 Factorise fully

If you wrote $4d^2 - 12d = 4(d^2 - 3d)$ you would not have factorised fully, because 4 is not the highest common factor of both terms.

(a) $4d^2 + 12d$

.... **(2 marks)**

(b) $6x^2 - 18x$

.... **(2 marks)**

(c) $7n^2 - 35n$

.... **(2 marks)**

Target grade 5

5 (a) Here are some factors.

$\boxed{2x^2 - 3xy \quad 3 \quad 3xy \quad 3x \quad 2x - 3y \quad 2x^2 - 3y}$

Write down the factors of $6x^2 - 9xy$ from the list.

Examiners' report Factorise $6x^2 - 9xy$. Each part outside the brackets, and the **whole** expression inside the brackets, are factors of the expression.

.... **(2 marks)**

(b) Write down all the factors of $12mn + 4m^2$

.... **(2 marks)**

Target grade 5

6 Show that when x is a whole number $5(2x + 3) - 4(x - 3)$ is a multiple of 3.

1. Expand the brackets.
2. Collect like terms.
3. Factorise.
4. Write a conclusion.

(2 marks)

29

ALGEBRA

Had a go ☐ Nearly there ☐ Nailed it! ☐

Linear equations 1

1 Solve

(a) $2x = 32$ $(\div 2)$

(b) $3x = -15$

(c) $-30 = 2q$

$x =$ .... **(1 mark)** $x =$ .... **(1 mark)** $q =$ .... **(1 mark)**

(d) $\dfrac{v}{-4} = 9$ $(\times -4)$

(e) $10 = \dfrac{x}{-12}$

(f) $72 = -8n$

$v =$ .... **(1 mark)** $x =$ .... **(1 mark)** $n =$ .... **(1 mark)**

2 Solve

(a) $x + 4 = 9$ (-4)

(b) $20 = p + 8$

(c) $7 - t = 5$

$x =$ .... **(1 mark)** $p =$ .... **(1 mark)** $t =$ .... **(1 mark)**

(d) $-2 = a - 5$

(e) $h - 4 = 15$

(f) $6 = -k - 40$

> The = sign is symmetrical. You can swap the right- and left-hand sides of the equation.

$a - 5 = -2$ $(+5)$

$a =$ .... **(1 mark)** $h =$ .... **(1 mark)** $k =$ .... **(1 mark)**

3 Solve

(a) $2x - 8 = 4$ $(+8)$

(b) $7p + 30 = 9$

$2x =$ .... $(\div 2)$

$x =$ .... **(2 marks)** $p =$ .... **(2 marks)**

(c) $4t + 25 = 9$

(d) $2 = 3f + 14$

$t =$ .... **(2 marks)** $f =$ .... **(2 marks)**

(e) $4 + \dfrac{h}{3} = 9$ (-4)

(f) $-2 = \dfrac{c}{3} + 6$

$\dfrac{h}{3} =$ .... $(\times 3)$

$h =$ .... **(2 marks)** $c =$ .... **(2 marks)**

4 Tom is planting some tulips. He buys 5 bags with t tulips in each bag. When he opens up the bags he finds 9 tulips are damaged.

(a) Write an expression for the number of tulips he can plant.

.... **(1 mark)**

(b) He planted 36 tulips. Form an equation in terms of t.

.... **(1 mark)**

(c) Solve the equation to find the number of tulips in each bag.

.... **(2 marks)**

Had a go ☐   Nearly there ☐   Nailed it! ☐   **ALGEBRA**

Linear equations 2

Target grade 3

Guided

5 Solve

(a) $6x + 3 = 2x + 11$ $(-2x)$

   $4x + 3 = 11$ (-3)

   $4x = \ldots\ldots$ $(\div 4)$

   $x = \ldots\ldots$

Collect the *x* terms on one side.

(3 marks)

(b) $5x + 4 = 3x - 12$

(c) $7t - 12 = 3t - 9$

$x = \ldots\ldots\ldots\ldots$ **(3 marks)** $t = \ldots\ldots\ldots\ldots$ **(3 marks)**

Target grade 4

6 Solve

Start by multiplying out the brackets.

(a) $3(2x - 1) = 27$

(b) $4(2x - 1) = 3x + 6$

$x = \ldots\ldots\ldots\ldots$ **(3 marks)** $x = \ldots\ldots\ldots\ldots$ **(3 marks)**

(c) $2(6 - x) = 3(2x + 12)$

(d) $\dfrac{4x + 8}{5} = 4$

$x = \ldots\ldots\ldots\ldots$ **(3 marks)** $x = \ldots\ldots\ldots\ldots$ **(3 marks)**

Target grade 4

7 The diagrams show two rectangular Christmas tags.

 $(4x + 1)$ cm    $(2x + 3)$ cm

4 cm   6 cm

Both tags have the same area.
Work out the value of x.

Examiners' report Don't use trial and improvement! Form an equation in *x* and solve it. Multiply the length by the width to find an expression for the area of each rectangle, then set these areas equal to each other.

$x = \ldots\ldots\ldots\ldots$ **(3 marks)**

Inequalities

1 Write down the inequalities shown on each number line.

> Closed (solid) circles show numbers **are** included.

(a)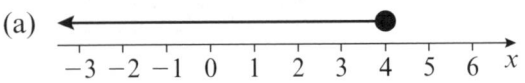

$x \leq$ .... **(1 mark)**

(b)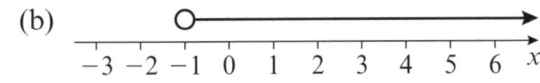

.... **(1 mark)**

(c)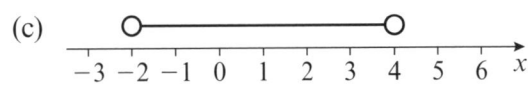

.... $< x <$ .... **(1 mark)**

(d)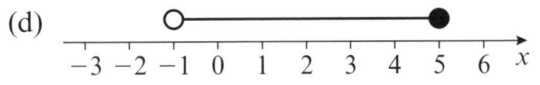

.... **(1 mark)**

2 Show each inequality **on** the number line.

> Use an open circle to show if a number is **not** included.

(a)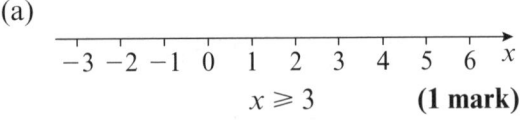

$x \geq 3$ **(1 mark)**

(b)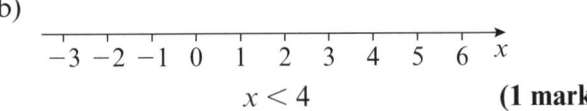

$x < 4$ **(1 mark)**

(c)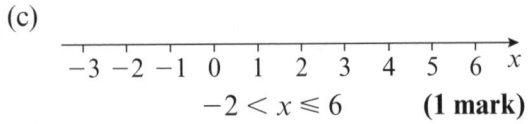

$-2 < x \leq 6$ **(1 mark)**

(d)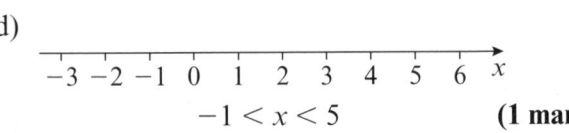

$-1 < x < 5$ **(1 mark)**

3 x is an integer. Write down all the possible values of x for each inequality.

> **Examiners' report** Don't include -3, and remember that 0 is an integer.

(a) $-3 < x \leq 4$

$x = -2,$ ...., ..., ..., ..., ..., ... **(1 mark)**

(b) $-2 \leq x < 3$

.... **(1 mark)**

(c) $-4 < x < 2$

.... **(1 mark)**

4 These numbers have been rounded. In each case, use inequalities to write an error interval for the number.

(a) $a = 180$ (to the nearest 10)

$175 \leq a <$ .... **(1 mark)**

(b) $b = 9000$ (to the nearest thousand)

.... **(1 mark)**

(c) $c = 123.5$ (to one decimal place)

.... **(1 mark)**

(d) $d = 10.8$ (to three significant figures)

.... **(1 mark)**

5 A hospital records the weight of a six-month-old baby as 7.83 kg, to 2 decimal places. Poppy and Jeff want to write an error interval for the weight of the baby, x kg.
Poppy writes $7.825 < x \leq 7.835$
Jeff writes $7.825 \leq x < 7.835$. Who is correct?

.... **(2 marks)**

Had a go ☐   Nearly there ☐   Nailed it! ☐   **ALGEBRA**

Solving inequalities

1 Solve

(a) $2x \leq 20$ $(\div 2)$

   $x \leq$ .... **(1 mark)**

(b) $3x > 15$

   .... **(1 mark)**

(c) $4x \geq 16$

   .... **(1 mark)**

(d) $3x \leq -16$

   $x \leq$ .... **(1 mark)**

(e) $5x - 10 > 0$

   .... **(1 mark)**

(f) $6x + 4 \geq 0$

   .... **(1 mark)**

2 Solve

(a) $3x + 1 \geq 19$ (-1)

   $3x \geq$ .... $(\div 3)$

   $x \geq$ .... **(2 marks)**

(b) $5x - 8 < 27$

   .... **(2 marks)**

(c) $5x - 12 > x$

   .... **(2 marks)**

(d) $4x + 6 \leq 2x$

   .... **(2 marks)**

3 x is an integer.
Write down all the possible values of x for each expression.

(a) $-6 < 2x \leq 2$

'Integer' means 'whole number'.

(b) $-5 < 3x \leq 13$

   .... **(3 marks)**

   .... **(3 marks)**

4 Find the integer value of x that satisfies both the inequalities.

   $2x - 3 > 5$ and $3x + 4 < 22$

Solve both inequalities first.

   .... **(3 marks)**

ALGEBRA

Had a go ☐ Nearly there ☐ Nailed it! ☐

Sequences 1

Target grade 1

1 Here are some patterns made from sticks.

Pattern number 1   Pattern number 2   Pattern number 3   Pattern number 4

Draw pattern number 4 in the space above. **(1 mark)**

Target grade 2 · Guided

2 Here are the first four terms of different number sequences. Write down the next two terms for each sequence.

(a) 2   6   10   14   ....   **(2 marks)**

(b) 3   8   13   18   ....   **(2 marks)**

(c) 1   3   9   27   ....   **(2 marks)**

(d) 1   4   9   16   ....   **(2 marks)**

Target grade 2

3 The rule for generating this sequence is 'add two consecutive terms to get the next term'.

1   4   7   ....

Work out the three missing numbers.

[1 + ... = 4]

(3 marks)

Target grade 3 · Guided

4 The terms in this sequence decrease by the same amount each time.

38   34   30   26

(a) Write down the next two terms in this sequence. .... **(2 marks)**

(b) Ravina says that 7 is a number in this sequence. Is she correct? Give a reason for your answer.

[Look at the numbers in the sequence. Do they have anything in common?]

.... **(2 marks)**

Target grade 3

5 Here is a sequence.
The third term of the sequence is 11.

....   11   ....

The rule for this sequence is
'add four to previous term then divide by two'.
Work out the first term of the sequence.

[**Problem solved!** Write the term-to-term rule as a function machine, then work backwards to find the second term, and then the first term.]

.... **(3 marks)**

34

Had a go ☐ Nearly there ☐ Nailed it! ☐   ALGEBRA

Sequences 2

 6 Here are some sequences. Write down an expression for the *n*th term of each linear sequence.

(a)
5 9 13 17
+4 +4 +4

*n*th term = 4*n* .... **(2 marks)**

(b) 2   5   8   11

.... **(2 marks)**

(c) 2   9   16   23

.... **(2 marks)**

(d) 8   13   18   23

.... **(2 marks)**

 7 Here are the first five terms of a linear sequence:

  4   7   10   13   16

Find an expression, in terms of *n*, for the *n*th term of the linear sequence.

.... **(2 marks)**

8 Here are some patterns made from sticks.

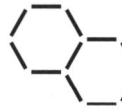   

Pattern   Pattern   Pattern   Pattern
number 1   number 2   number 3   number 4

(a) Draw pattern number 4 in the space above.   **(1 mark)**

Copy pattern number 3 and then draw in some more sticks to make pattern number 4.

(b) Write down a formula for the number of sticks, *S*, in terms of the pattern number, *n*.

S = ... **(2 marks)**

 9 Here are the first four terms of an arithmetic sequence:

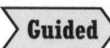
3   7   11   15   19

(a) Find an expression, in terms of *n*, for the *n*th term of the sequence.

.... *n* − .... **(2 marks)**

(b) Molly says that 199 is a term in the arithmetic sequence. Is Molly correct? Give a reason for your answer.

 **Problem solved!** Set the *n*th term of the sequence equal to 199 and solve the equation to find *n*. If *n* is an integer then the term is part of the sequence. If *n* is not an integer then the term is **not** part of the sequence.

.... **(2 marks)**

35

ALGEBRA   Had a go ☐   Nearly there ☐   Nailed it! ☐

Coordinates

Target grade 1

Guided

1 (a) Write down the coordinates of

   (i) point A

   (4, ...)   **(1 mark)**

   (ii) point B

   (.........., ...)   **(1 mark)**

   Remember the first number is the distance from the origin left or right, the second number is the distance from the origin up or down.

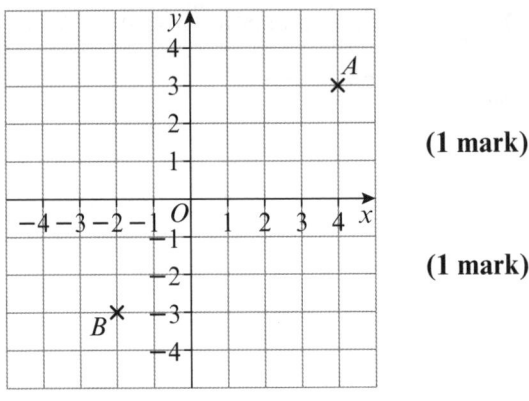

  (b) On the grid, plot the point
   (i) (4, −2) and label it C   **(1 mark)**
   (ii) (−1, 3) and label it D.   **(1 mark)**

Target grade 3

Guided

2 Work out the mid-points for the line segments given by the following pairs of coordinates.

   (a) (3, 6) and (7, 12)

   Mid-point = $\left(\dfrac{3+7}{2}, \dfrac{\text{........} + \text{........}}{2}\right) = (\text{..........}, \text{..........})$ **(2 marks)**

   (b) (1, 8) and (9, 3)   (c) (4, 7) and (−8, 13)   (d) (2, −6) and (−10, 12)

   ....
   (2 marks)   **(2 marks)**   **(2 marks)**

Target grade 3

3 Work out the mid-points of the following line segments.

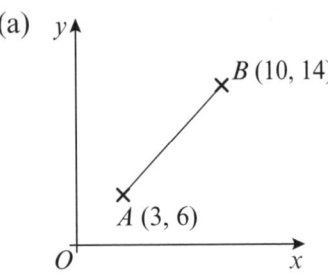

   (a)   (b)

   ....   **(2 marks)**

Target grade 3

4 Two straight lines are shown.
   A is the mid-point of OB and B is the mid-point of PQ.
   Show that the coordinates of P are (2, 10)

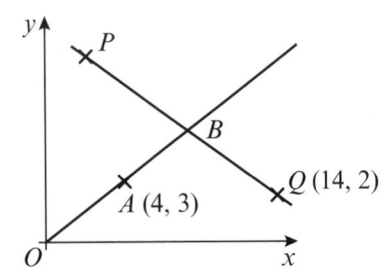

Problem solved! Start by working out the coordinates of B. Don't assume that P has coordinates (2, 10). You need to use the other information given to show this.

(2 marks)

Had a go ☐   Nearly there ☐   Nailed it! ☐   ALGEBRA

Gradients of lines

Target grade 3

Guided

1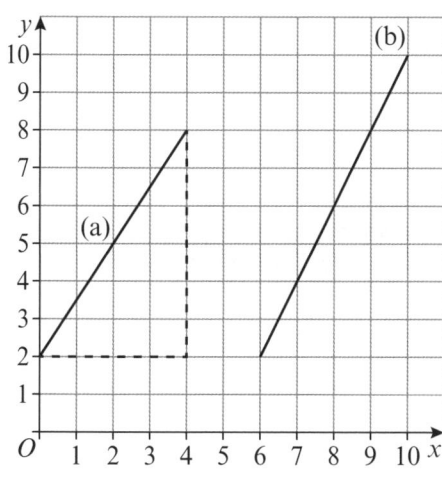

Work out the gradients of the straight lines shown on the grid.

> Draw a triangle on the graph and use it to find the gradient.

(a) Gradient = $\dfrac{\text{distance up}}{\text{distance across}}$

= $\dfrac{............}{............}$

= ....   **(2 marks)**

(b)

.... **(2 marks)**

Target grade 3

Guided

2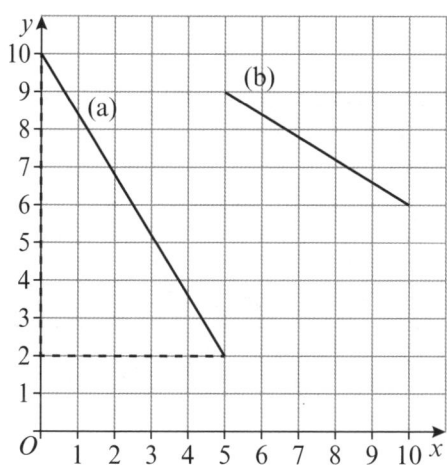

Work out the gradients of the straight lines shown on the grid.

> The lines slope **down** so the gradient is **negative**.

(a) Gradient = $\dfrac{\text{distance up}}{\text{distance across}}$

= $\dfrac{............}{............}$

= ....   **(2 marks)**

(b)

.... **(2 marks)**

Target grade 3

3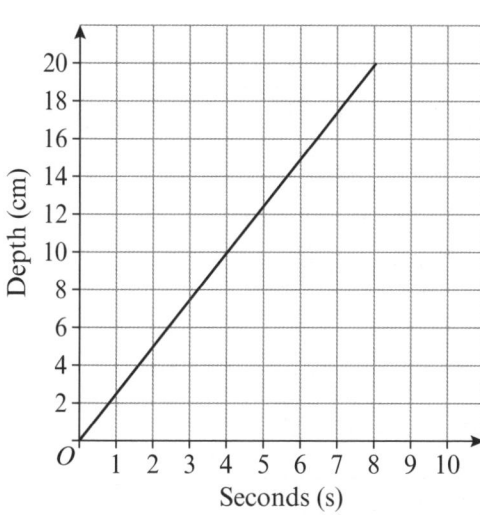

Kim pours diesel into a container. The graph shows how the depth, in cm, of the diesel changes with time, in seconds.
Show that the rate of change of the depth of the diesel is 2.5 cm/s.

> Look carefully at the vertical scale.

.... **(2 marks)**

Straight-line graphs 1

1

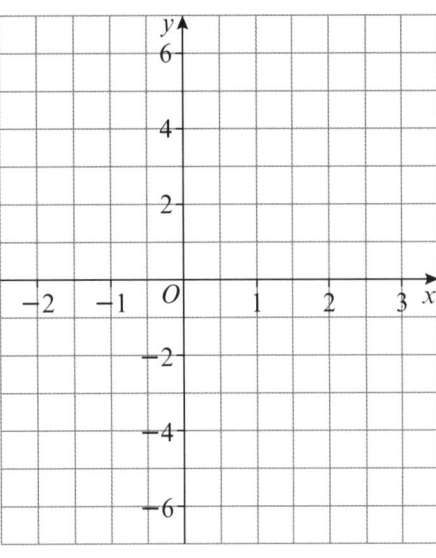

x	-2	-1	0	1	2	3
y		-3				5

(a) Complete the table of values for $y = 2x - 1$

> Substitute each value for x into the rule $y = 2x - 1$ to find the value of y.

$x = -2$: $y = (2 \times -2) - 1$

$= \ldots\ldots - 1 = \ldots\ldots$

$x = 1$: $y = (2 \times \ldots\ldots) - 1$

$= \ldots\ldots - 1 = \ldots\ldots$ **(2 marks)**

(b) On the grid draw the graph of $y = 2x - 1$

(2 marks)

2

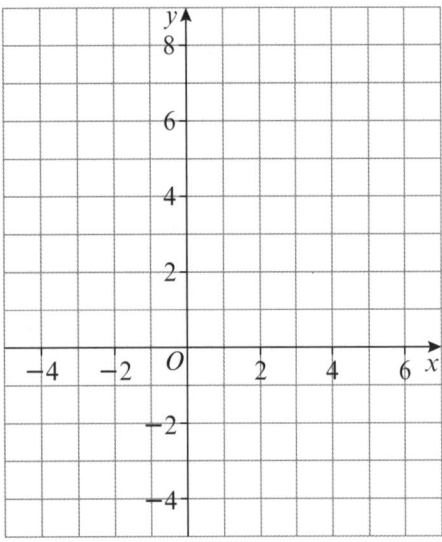

On the grid draw the graph of $x + y = 5$ for the values of x from -3 to 6.

> First draw a table of values. The question tells you to use 'values of x from -3 to 6'. Next work out the values of y.

(2 marks)

3

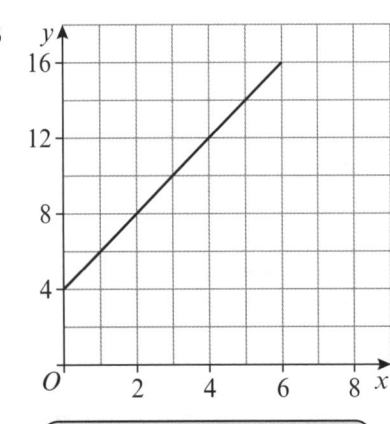

Find the equation of the straight line.

> Draw a triangle on the graph and use it to find the gradient.

$$\text{Gradient} = \frac{\text{distance up}}{\text{distance across}} = \frac{\ldots\ldots\ldots}{\ldots\ldots\ldots}$$

$= \ldots\ldots\ldots$

$y = \ldots\ldots + \ldots\ldots$ **(2 marks)**

> Use $y = mx + c$ to find the equation of a straight line.

> c is the intercept on the y-axis.

Had a go ☐ Nearly there ☐ Nailed it! ☐ ALGEBRA

Straight-line graphs 2

 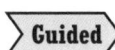

4 Find the equation of the straight line with

(a) gradient 3, passing through the point (2, 5)

$y = 3x + c$

$5 = \ldots \times \ldots + c$

$c = \ldots$

$y = \ldots x - \ldots$ **(2 marks)**

(b) gradient −2, passing through the point (3, 6)

.... **(2 marks)**

> Substitute the value of the gradient into $y = mx + c$. Then substitute the x-values and y-values given into your equation. Solve the equation to find c. Remember to write your completed equation at the end.

(c) gradient 4, passing through the point (−2, 7)

.... **(2 marks)**

(d) gradient 4, passing through the point (−1, −6).

.... **(2 marks)**

 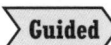

5 Find the equation of a straight line which passes through the following points.

(a) (3, 2) and (5, 6)

$m = \ldots$

$2 = \ldots \times \ldots + c$

$c = \ldots$

$y = \ldots x - \ldots$ **(3 marks)**

(b) (−1, 5) and (4, −10)

.... **(3 marks)**

6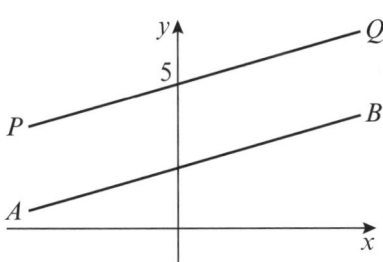

Here are two straight lines.
The equation of line AB is $y = 4x + 1$
Line AB is parallel to line PQ.
Find the equation of line PQ.

>  Find the gradient of AB and the value of the y-intercept for PQ.
> Then use $y = mx + c$ to find the equation of line PQ.

.... **(2 marks)**

39

Real-life graphs

1 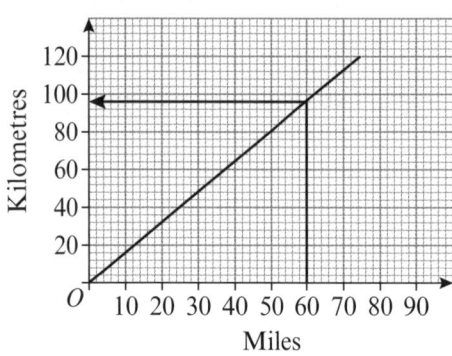 You can use this graph to change between miles and kilometres.

(a) Use the graph to change 60 miles into kilometres.

60 miles = ... kilometres **(1 mark)**

> Draw a vertical line from 60 miles to the line and then draw a horizontal line across to the kilometres.

(b) The distance from Rome to Lyon is 660 miles.
The distance from Rome to Marseille is 950 kilometres.
Is Rome closer to Lyon or closer to Marseille?
You must show all of your working.

> The horizontal scale on the graph does not go up to 660 miles. Use your answer to part (a) to work out 660 miles in kilometres, then write a conclusion.

(3 marks)

2 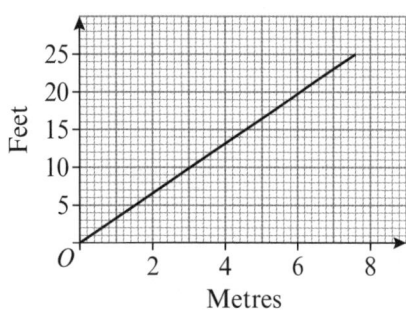 You can use this graph to change between feet and metres.

(a) Use the graph to change 15 feet into metres.

..........................m **(1 mark)**

Amy and Sandeep are throwing a shot put.

(b) Amy throws the shot put 18 feet and Sandeep throws it 6 metres.
Who throws the shot put the furthest? You must show all of your working.

(3 marks)

3 Lisa lays lawns in gardens of different areas.
She uses this graph to work out the cost of laying the lawn.

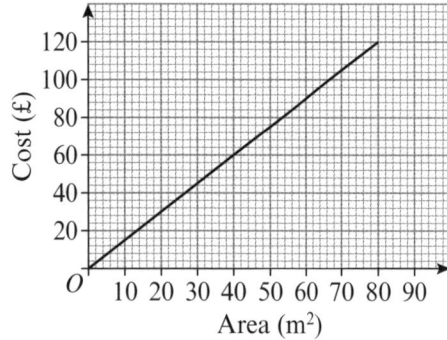

(a) She lays down 40 m² of lawn.
Use the graph to find the cost of laying the lawn.

£.......................... **(1 mark)**

(b) Lisa says, 'My price increases by £1.50 for every square metre.'
Is there evidence available to support this statement?

.... **(3 marks)**

Had a go ☐ Nearly there ☐ Nailed it! ☐   **ALGEBRA**

Distance–time graphs

Guided

1 Becky cycled from her home to the shop. She went into the shop. She then cycled back home. Here is a distance–time graph for Becky's complete journey.

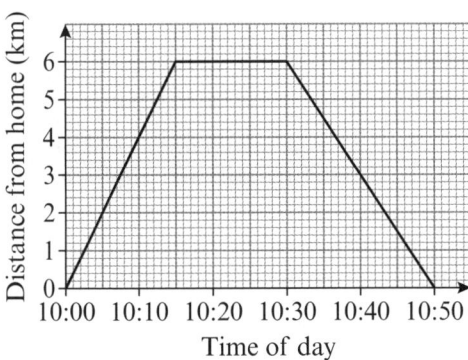

(a) What time did Becky start her journey?

Start time =

.... **(1 mark)**

(b) What is the distance from Becky's home to the shop?

After how many kilometres does she stop?

.... km **(1 mark)**

(c) How many minutes was Becky in the shop? *This is where the graph is horizontal.*

.... **(1 mark)**

(d) Work out Becky's average speed for her return journey.

speed = $\frac{\text{distance}}{\text{time}}$ = $\frac{\text{...............}}{\text{...............}}$ = ... km/h   **(2 marks)**

2 Gary left home at 1 pm to go for a walk. The distance–time graph represents part of Gary's journey.

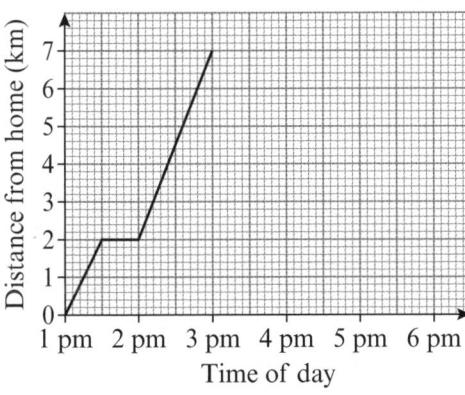

(a) Gary stopped for a break at 1.30 pm. Write down how many minutes Gary stopped for.

.... **(1 mark)**

(b) How far was Gary from home at 2 pm?

Draw a line up from 2 pm and then across to the vertical axis.

.... **(1 mark)**

Gary had a rest at 3 pm for one hour. He then walked home at a steady speed. His walk home took him one and a half hours.

(c) Complete the distance–time graph.   **(2 marks)**

3 Nisha drives 40 km from her home to her grandparents' home. The journey took one hour. She spent three hours at their house, and then drove home at a steady speed. Her journey home took 30 minutes.

(a) Draw a distance–time graph of Nisha's trip.

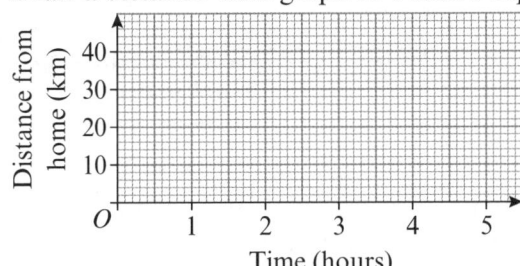

(2 marks)

(b) Work out Nisha's average speed for her return journey.

.... km/h **(2 marks)**

41

ALGEBRA

Had a go ☐   Nearly there ☐   Nailed it! ☐

Rates of change

Target grade 3

1. Here are four flasks. Rachael fills each flask with water.
   The graphs show the rate of change of the depth of the water in each flask as Rachael fills it.
   Draw a line from each flask to the correct graph. One line has been drawn for you.

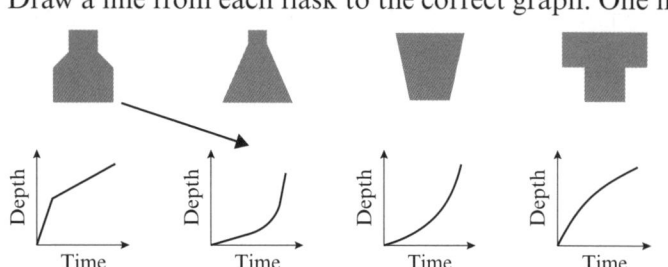

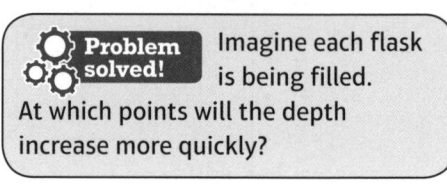

Problem solved! Imagine each flask is being filled. At which points will the depth increase more quickly?

(2 marks)

Target grade 3

2. The graph shows how much money there was in Dan's savings account over the last 12 months.

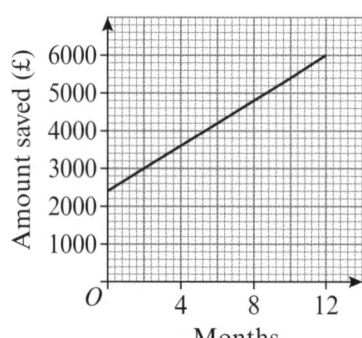

   (a) How much money was there at the start?

   £.......................... **(1 mark)**

   (b) Work out the gradient of the line.

   Draw a triangle on the graph and use it to find the gradient.

   .... **(2 marks)**

   (c) Interpret the value of the gradient.

   .... **(1 mark)**

Target grade 5

Guided

3. Here is a velocity–time graph of a car.

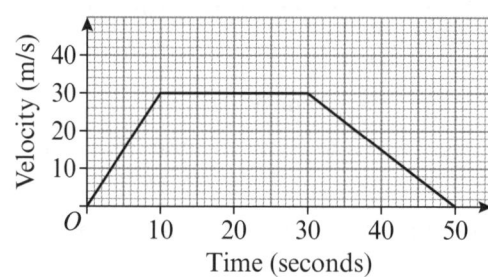

Draw a triangle on the graph and use it to find the gradient.

   (a) What is the rate of change of velocity in the first 10 seconds?

   Gradient = $\dfrac{\text{distance up}}{\text{distance across}}$ = $\dfrac{\ldots\ldots}{\ldots\ldots}$ = ... m/s²

   (2 marks)

   (b) Describe what is happening to the car between 10 seconds and 30 seconds.

   This is where the graph is horizontal.

   .... **(1 mark)**

   (c) What is the rate of change of velocity in the last 20 seconds?

   .... m/s² **(2 marks)**

Had a go ☐ Nearly there ☐ Nailed it! ☐   ALGEBRA

Expanding double brackets

 1 Expand and simplify   'Expand' means 'multiply out the brackets'.

   (a) $(x + 3)(x + 5)$

   $= x(x + 5) + 3(x + 5)$

   $= x^2 + \ldots\ldots\ldots + \ldots\ldots\ldots + \ldots\ldots\ldots$

   This is the 'one at a time' method. You can use any method you like to expand the brackets.

   $= x^2 + \ldots\ldots\ldots + \ldots\ldots\ldots$ (2 marks)

   (b) $(x + 3)(x + 2)$ (c) $(x + 1)(x + 4)$

   .... (2 marks)   .... (2 marks)

   (d) $(x + 2)(x - 5)$ (e) $(x - 2)(x - 5)$

   .... (2 marks)   .... (2 marks)

 2 Expand and simplify

   (a) $(3x - 4)(5x - 1)$

   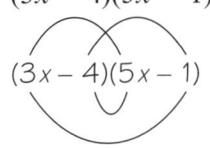

   This is the 'FOIL' method. Multiply the First terms, then the Outside terms, then the Inside terms, then the Last terms.

   $= 15x^2 - \ldots\ldots\ldots\ldots - \ldots\ldots\ldots\ldots + \ldots\ldots\ldots\ldots$

   $= 15x^2 - \ldots\ldots\ldots\ldots + \ldots\ldots\ldots\ldots$ (2 marks)

   (b) $(8x - 3)(2x - 1)$ (c) $(7x - 5)(4x - 5)$

   .... (2 marks)   .... (2 marks)

   (d) $(x + 3)^2$ (e) $(2x - 5)^2$

   .... (2 marks)   .... (2 marks)

 3 n is a whole number. Show that $(n + 1)^2 - n^2$ is always an odd number.

   .... (2 marks)    Odd numbers can be written as $2 \times$ whole number $+ 1$.

 4 x and y are integers. Show that $x^2 + y^2 - (x - y)^2$ is always an even number.

   (2 marks)

 5 The diagram shows a rectangle with length $(2x + 3)$ cm and width $(x - 2)$ cm. The area of the rectangle is A square centimetres. Show that $A = 2x^2 - x - 6$

   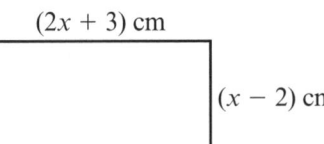

   (2 marks)

43

Quadratic graphs

ALGEBRA — Had a go ☐ Nearly there ☐ Nailed it! ☐

Target grade 4

1 (a) Complete the table of values for $y = x^2 - 2$

Guided

x	−3	−2	−1	0	1	2	3
y		2					7

Substitute each value of x into the formula $y = x^2 - 2$ to find the value of y.

$x = -3$: $y = (-3 \times -3) - 2$
$= \ldots\ldots\ldots - 2$
$= \ldots\ldots\ldots$

$x = 1$: $y = (1 \times \ldots\ldots\ldots) - 2$
$= \ldots\ldots\ldots$

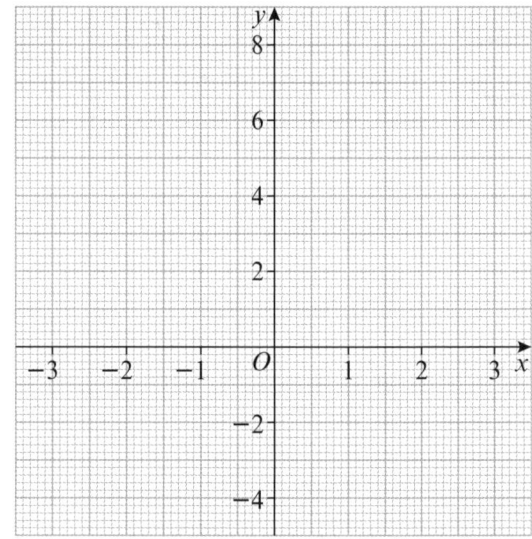

(2 marks)

(b) On the grid draw the graph of $y = x^2 - 2$ **(2 marks)**

(c) Write down the coordinates of the turning point.

The turning point is the point where the direction of the curve changes.

………………………………… **(1 mark)**

(d) Use your graph to find the value of y when $x = 2.5$ …………… **(1 mark)**

Target grade 4

2 (a) Complete the table of values for $y = x^2 - 4x + 3$

Guided

x	−1	0	1	2	3	4	5
y			0				

Substitute each value for x into the rule $y = x^2 - 4x + 3$ to find the value of y.

$x = -1$: $y = (-1 \times -1) - (4 \times -1) + 3 = \ldots\ldots\ldots\ldots$

$x = 3$: $y = (3 \times \ldots\ldots\ldots) - (4 \times \ldots\ldots\ldots) + 3$

$= \ldots\ldots\ldots$ **(2 marks)**

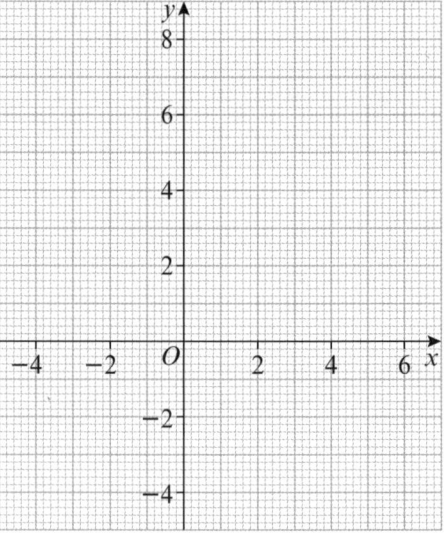

(b) On the grid draw the graph of $y = x^2 - 4x + 3$ **(2 marks)**

(c) Write down the coordinates of the turning point.

………………………………… **(2 marks)**

Had a go ☐   Nearly there ☐   Nailed it! ☐

ALGEBRA

Using quadratic graphs

Target grade 4

Guided

1 (a) Complete the table of values for
   $y = x^2 - x - 4$

x	−3	−2	−1	0	1	2	3
y				−4			

$x = -2$: $y = (-2 \times -2) - (-2) - 4$

= ....

$x = 1$: $y = (1 \times 1) - (1) - 4$

= ....   **(2 marks)**

(b) On the grid draw the graph of
   $y = x^2 - x - 4$ **(2 marks)**

(c) Use your graph to write down an estimate for

   (i) the minimum value of y

   > The minimum value of y is the point where the direction of the curve changes.

   .... **(1 mark)**

   (ii) the solutions of $x^2 - x - 4 = 0$

   .... **(2 marks)**

(d) Comment on the accuracy of your estimates.

.... **(1 mark)**

Target grade 2

Guided

2 (a) Complete the table of values for $y = 8 + 3x - 2x^2$

x	−2	−1	0	1	2	3
y		3				

$x = -2$: $y = 8 + (3 \times -2) - (2 \times -2 \times -2)$

= ....

$x = 3$: $y = 8 + (3 \times 3) - (2 \times 3 \times 3)$

= ....   **(2 marks)**

(b) On the grid draw the graph of $y = 8 + 3x - 2x^2$ **(2 marks)**

(c) Write down the coordinates of the turning point.

   .... **(2 marks)**

> **Examiners' report**
> Remember to join points on a quadratic graph with a **smooth curve**.

ALGEBRA

Had a go ☐ Nearly there ☐ Nailed it! ☐

Factorising quadratics

1 Factorise

(a) $x^2 + 4x + 3$

.... × .... = +3, ... + ... = +4

Examiners' report: The answer will have two sets of brackets. You can check your answer by multiplying out the brackets and simplifying.

$x^2 + 4x + 3 = (x + ...)(x + ...)$ **(2 marks)**

You need to find two numbers that multiply to give 3 and add up to give 4.

(b) $x^2 + 11x + 10$

(..........) × (..........) = +10   (..........) + (..........) = +11

$x^2 + 11x + 10 = (x ....)(x ....)$ **(2 marks)**

(c) $x^2 + 6x + 5$ (d) $x^2 - 11x + 10$ (e) $x^2 - 12x + 20$

.... **(2 marks)**   .... **(2 marks)**   .... **(2 marks)**

(f) $x^2 - 9x + 14$

(..........) × (..........) = +14   (..........) + (..........) = −9

$x^2 - 9x + 14 = (x ....)(x ....)$ **(2 marks)**

2 Factorise

(a) $x^2 + 6x - 7$ (b) $x^2 + 4x - 5$ (c) $x^2 - 2x - 15$

.... **(2 marks)**   .... **(2 marks)**   .... **(2 marks)**

3 Factorise

(a) $x^2 - 13x + 22$ (b) $x^2 - 6x - 16$ (c) $x^2 - 14x + 40$

.... **(2 marks)**   .... **(2 marks)**   .... **(2 marks)**

4 Factorise

(a) $x^2 - 9$
    $a = x, b = 3$

This is a difference of two squares. You can use the rule $a^2 - b^2 = (a + b)(a - b)$

    $x^2 - 9 = (x + ...)(x - ....)$ **(2 marks)**

(b) $x^2 - 144$ (c) $x^2 - 81$ (d) $x^2 - 64$

.... **(2 marks)**   .... **(2 marks)**   .... **(2 marks)**

(e) $x^2 - 1$ (f) $x^2 - 169$

.... **(2 marks)**   .... **(2 marks)**

Had a go ☐   Nearly there ☐   Nailed it! ☐   **ALGEBRA**

Tricky Topic

Quadratic equations

1 Solve

(a) $x^2 - 3x = 0$

.... ($x -$ ....) $= 0$

$x = 0$ or $x =$ .... **(2 marks)**

> Find the values of x that make each factor equal to 0. The first factor is just x so one solution is $x = 0$

(b) $x^2 + 5x = 0$ .... **(2 marks)**

(c) $x^2 - 7x = 0$ .... **(2 marks)**

2 Solve

(a) $x^2 + 6x + 8 = 0$

$(x + 2)(x +$ ....$) = 0$

$x = -2$ or $x =$ ....

> The first factor is $x + 2$, so the first solution is $x = -2$

(2 marks)

(b) $x^2 - 7x + 12 = 0$ .... **(2 marks)**

(c) $x^2 + 9x + 20 = 0$ .... **(2 marks)**

(d) $x^2 + 8x + 7 = 0$ .... **(2 marks)**

(e) $x^2 - 2x - 24 = 0$

(x ....)(x ....) $= 0$

$x =$ .... or $x =$ .... **(2 marks)**

3 Solve

> Use the rule for the difference of two squares: $a^2 - b^2 = (a + b)(a - b)$

(a) $x^2 - 4 = 0$

$(x +$ ....)($x -$ ....) $= 0$

$x =$ .... or $x =$ .... **(2 marks)**

(b) $x^2 - 25 = 0$ .... **(2 marks)**

(c) $x^2 - 49 = 0$ .... **(2 marks)**

(d) $x^2 - 121 = 0$ .... **(2 marks)**

(e) $x^2 - 9 = 0$ .... **(2 marks)**

4 The sum of the squares of two consecutive positive numbers is 221. Find the two positive numbers.

> **Problem solved!** Written using algebra, this is $n^2 + (n + 1)^2 = 221$

.... **(3 marks)**

ALGEBRA

Had a go ☐ Nearly there ☐ Nailed it! ☐

Cubic and reciprocal graphs

1 (a) Complete the table of values for $y = x^3 - 4x - 2$

x	-2	-1	0	1	2	3
y		1				13

Substitute each value for x into the rule $y = x^3 - 4x - 2$ to find the value of y.

$x = -2$: $y = (-2 \times -2 \times -2)$
$- (4 \times -2) - 2 = $ ....

$x = 1$: $y = (1 \times 1 \times 1) - (4 \times 1) - 2$

$= $ .... **(2 marks)**

(b) On the grid draw the graph of $y = x^3 - 4x - 2$ **(2 marks)**

(c) Write down the coordinates of the turning points.

.... **(2 marks)**

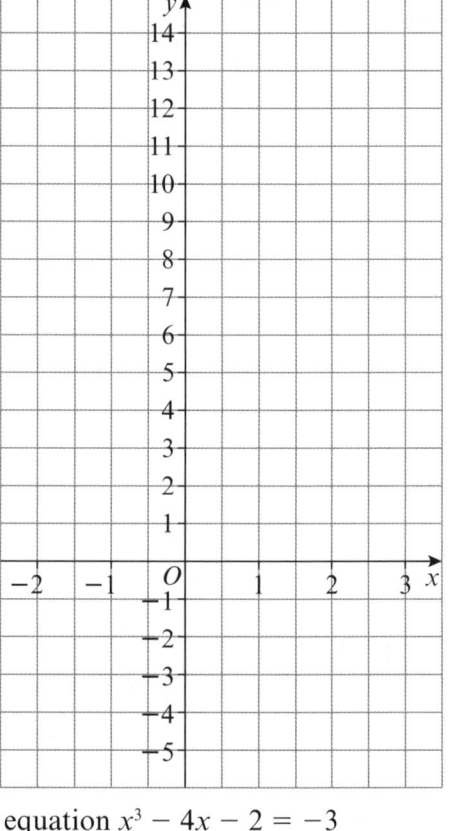

(d) Use your graph to find estimates of the solutions to the equation $x^3 - 4x - 2 = -3$

Draw the line $y = -3$ on the graph. Find the x-coordinates of the points of intersection with the curve.

.... **(2 marks)**

2   A   B   C   D   E

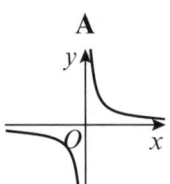 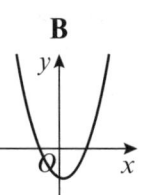 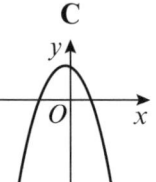 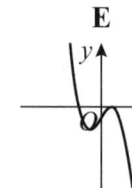

Write down the letter of the graph which could have the equation

(i) $y = x^2 - x - 6$ (ii) $y = x^3 - 3x + 5$ (iii) $y = \dfrac{1}{x}$

.... **(1 mark)**   .... **(1 mark)**   .... **(1 mark)**

(iv) $y = 6 - x - x^2$ (v) $y = 2 + 3x - x^3$

.... **(1 mark)**   .... **(1 mark)**

48

Had a go ☐   Nearly there ☐   Nailed it! ☐   **ALGEBRA**

Simultaneous equations

1 Solve the simultaneous equations

| Label the equations (1) and (2). |

(a) $2x + 5y = 16$ (1)
    $5x - 2y = 11$ (2)

(1) × 5 gives   .... x + ... y = ... (3)
(2) × 2 gives $10x$ $- 4y = 22$ (4)
(3) − (4) gives

.... y = ...

$y = $ ....

Substitute $y = $ .... in (1)

$2x + 5 \times$ .... $= 16$

$x = $ ....

$x = $ .... $y = $ .... **(3 marks)**

(b) $3x + 2y = 11$
    $2x - 5y = 20$

$x = $ .... $y = $ .... **(3 marks)**

2 By drawing two suitable straight lines on the coordinate grid below, solve the simultaneous equations

(a) $x + y = 5$
    $y = 3x + 1$

(b) $2x + y = 5$
    $x + y = 3$

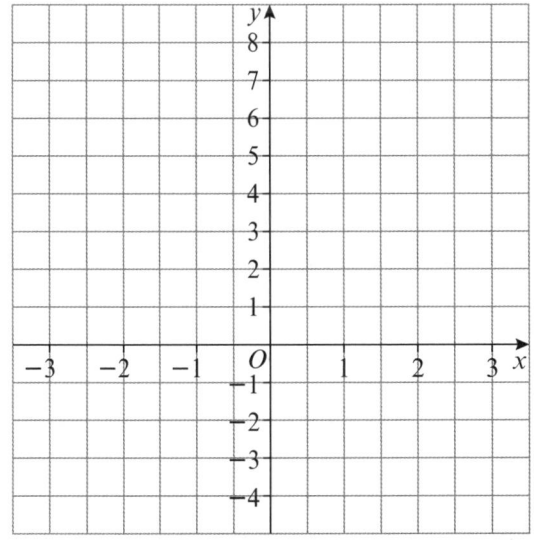
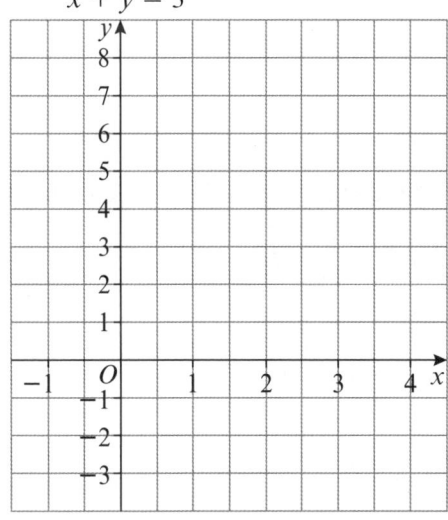

For each equation choose three x-values and then find the corresponding y-values.
Plot these points and draw a straight line through the three points.
The solution to the simultaneous equations is the point where the lines cross.

(a) $x = $ .... $y = $ .... **(4 marks)**   (b) $x = $ .... $y = $ .... **(4 marks)**

ALGEBRA

Had a go ☐   Nearly there ☐   Nailed it! ☐

Rearranging formulae

1 A straight line has equation $3y = 4x + 3$. The point P lies on the straight line. P has a y-coordinate of 5. Work out the x-coordinate of P.

> Substitute $y = 5$ into $3y = 4x + 3$ and solve the equation to find x.

$3 \times \ldots\ldots = 4x + 3$

$\ldots\ldots - \ldots\ldots = 4x$

$x = \ldots\ldots \div \ldots\ldots = \ldots\ldots$ **(2 marks)**

2 Mandy wants to work out the time in seconds taken for a ball to reach the ground by using the formula $v = u + at$. She knows that $v = 30$, $u = 5$ and $a = 10$. Work out the time taken, t, for the ball to reach the ground.

$t = \ldots\ldots$ **(3 marks)**

3 Make the letter in the brackets the subject of each formula.

(a) $v = u + 10t$ (t)

$v - \ldots\ldots = 10t$ $(\div 10)$

$\ldots\ldots = t$

$t = \ldots\ldots$ **(2 marks)**

(b) $m = 6n + 19$ (n)

$n = \ldots\ldots$ **(2 marks)**

(c) $d = ut + at^2$ (u)

$u = \ldots\ldots$ **(2 marks)**

(d) $P = A - 6D$ (D)

$D = \ldots\ldots$ **(2 marks)**

4 Make the letter in the brackets the subject of each formula.

(a) $s = \dfrac{d}{t}$ (t)

$t = \ldots\ldots$ **(2 marks)**

(b) $d = \sqrt{\dfrac{5h}{4}}$ (h)

$h = \ldots\ldots$ **(2 marks)**

(c) $s = \tfrac{1}{2}(u + v)t$ (t)

$t = \ldots\ldots$ **(2 marks)**

(d) $v^2 = u^2 + 2as$ (s)

$s = \ldots\ldots$ **(2 marks)**

5 Make the letter in the brackets the subject of each formula.

(a) $P = h(2 + n)$ (n)

$n = \ldots\ldots$ **(3 marks)**

(b) $t = 3(1 - 2x)$ (x)

$x = \ldots\ldots$ **(3 marks)**

Had a go ☐   Nearly there ☐   Nailed it! ☐   **ALGEBRA**

Using algebra

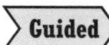

1. Tom plants some rhubarb seeds in his allotment.
   He plants 7 rows of these seeds. In each row there are x seeds.
   A few months later he finds insects have eaten 9 seeds.
   After this he has 96 rhubarb plants left in total.

   (a) Write down an equation using this information.

   $7 \times$ .... $-$ .... $=$ .... **(1 mark)**

   (b) Work out the value of x.

   [Solve the equation.]

   .... **(2 marks)**

2. The diagram shows a triangular playground.
   The total perimeter of the playground is 54 m.
   Work out the value of x.

   [**Problem solved!** Form an equation by adding together the lengths of each side, then setting this equal to 54. Simplify your equation and solve it to find x.]

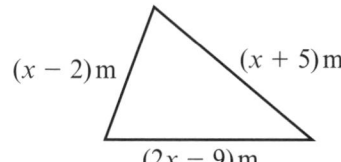

$(x - 2) +$ .... $+$ .... $= 54$

$x =$ .... **(3 marks)**

3. The diagram shows two rectangular gardens. All the measurements are in metres.

   $3x - 2$ [A] $2x + 3$ $2x + 1$ [B] $2x + 6$

   Both gardens have the same perimeter. Work out the width and height of garden A.

   .... **(3 marks)**

4. A square has side length $2x$ cm.
   An equilateral triangle has side length $(2x + 4)$ cm.
   The perimeter of the square is equal to the perimeter of the equilateral triangle.
   Work out the value of x.

   $x =$ .... **(3 marks)**

ALGEBRA — Had a go ☐ Nearly there ☐ Nailed it! ☐

Identities and proof

Tricky Topic

Target grade 5 — **Guided**

1. Show that $(2n - 1)^2 + (2n + 1)^2 \equiv 2(4n^2 + 1)$

   Multiply out the brackets.

   $(2n - 1)^2 = (2n - 1)(2n - 1) = $ ....

   $(2n + 1)^2 = (2n + 1)(2n + 1) = $ ....

   Add the brackets together and then factorise.

   (2 marks)

Target grade 5 — **Guided**

2. Show that the sum of any four consecutive integers is always a multiple of 2.

   You need to use algebra – it's not enough to just try this with four numbers.

   You can represent any four consecutive integers as n, $n + 1$, and so on.

   $n + (n + 1) + $ .... $ + $ ....

   $ = $ .... $n + $ ....

   $ = $ .... (.......... $n + $ ....)

   .... is a factor so the total expression must be a multiple of ....   **(3 marks)**

Target grade 5

3. Given that $5(x - c) = 4x - 5$ where c is an integer, show that x is a multiple of 5.

   Multiply out the brackets and then rearrange to make x the subject.

   (3 marks)

Target grade 5

4. Show that
   (a) $(x - 1)^2 \equiv x^2 - 2x + 1$
   (b) Hence, or otherwise, show that $(x + 1)^2 + (x - 1)^2 \equiv 2(x^2 + 1)$

   (2 marks)   **(2 marks)**

Target grade 5

5. (a) Show that the sum of any three **consecutive** even numbers is always a multiple of 6.

   (b) Give an example to show that the sum of any three even numbers is not necessarily a multiple of 6.

   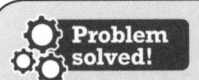 Write your even numbers as $2n$, $2n + 2$ and $2n + 4$.

   You don't need to use algebra for part (b). Just find three even numbers whose sum is not a multiple of 6.

   (3 marks)   ....   **(1 mark)**

52

Had a go ☐ Nearly there ☐ Nailed it! ☐   **ALGEBRA**

Problem-solving practice 1

Target grade 1

1 Tom cleaned his swimming pool. He hired a cleaning machine to do this job.
The cost of hiring the cleaning machine was £35.50 for the first day and then £18.25 for each extra day.
Tom's total cost of hiring the machine was £163.25.
For how many days did Tom hire the machine?

.... days **(3 marks)**

Target grade 3

2 The coordinates of three vertices of a square are (2, 1), (2, 5) and (6, 5).
Write down the coordinates of the

(a) missing vertex   (b) centre of the square.

.... **(1 mark)**   .... **(2 marks)**

Target grade 3

3 A gardener charges £15 for each hour he works at a job plus £25. The cost, in £, of the job can be worked out using the formula
    Cost = number of hours worked × 15 + 25

(a) The gardener works 7 hours. Work out the total cost.

£.................................. **(2 marks)**

(b) He charges £115 for one job. How many hours did he work?

.... hours **(2 marks)**

Target grade 3

4 Dan and Jay are measuring the distance a ball is rolled.
They use the formula $s = ut + 5t^2$ where $u = 4$ and $t = 2$
Dan works out the value of s to be 28 and Jay works out the value of s to be 108.
Who is correct? Give a reason for your answer.

.... **(3 marks)**

Target grade 4

5 Here are some patterns made using sticks.

Anjali says that for pattern number 10 there will be 32 sticks. Is she correct?

.... **(1 mark)**

53

ALGEBRA

Had a go ☐ Nearly there ☐ Nailed it! ☐

Problem-solving practice 2

Target grade 3

6 The diagram shows a square.
   Find the length of one side of the square.

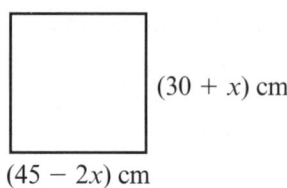

.... cm **(2 marks)**

Target grade 3

7 ABC is a triangle. Work out the size of the smallest angle.

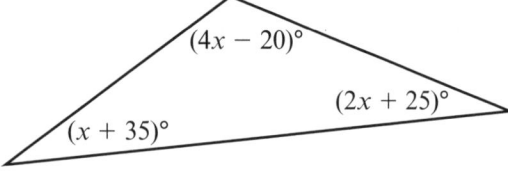

....................................° **(3 marks)**

Target grade 5

8 (a) Write down the equation of a straight line that is parallel to $y = 5x + 4$

.... **(1 mark)**

(b) Write down the equation of the straight line that is parallel to $y = 3x + 5$ and which passes through $(4, 7)$.

.... **(2 marks)**

Target grade 4

9 Here are the first five terms of a linear sequence:

   2   5   8   11   14

34 cannot be a term in this linear sequence. Explain why.

.... **(3 marks)**

Target grade 5

10 $(3x + 4)$ cm

    $(2x - 1)$ cm

The diagram shows a rectangle with length $(3x + 4)$ cm and width $(2x - 1)$ cm.
The area of the rectangle is A square centimetres.
Show that $A = 6x^2 + 5x - 4$

(2 marks)

54

Had a go ☐   Nearly there ☐   Nailed it! ☐   **RATIO & PROPORTION**

Percentages

Target grade 1

Guided

1 Find

(a) 9% of 50

(b) 4% of 275.

$\frac{9}{100} \times$ .... = ...

(2 marks)

.... **(2 marks)**

Target grade 2

Guided

2 Express the following as percentages:

(a) 35 out of 56

(b) 15 out of 75.

$\frac{35}{56} \times$ .... = ... %

(2 marks)

..............................% **(2 marks)**

Target grade 1

3 A ticket for the theatre costs £96 plus a booking charge of 8%. Work out

> For part (a), find 8% of £96. For part (b), add this amount on to the original price.

(a) the amount of the booking charge

(b) the total cost of a theatre ticket.

£.............................. **(2 marks)**

£.............................. **(2 marks)**

Target grade 2

4 Alice bought a car for £15 000. After one year the car was worth 12% less. Work out the new value of the car.

£.............................. **(3 marks)**

Target grade 2

5 Suki invited 128 people to a Christmas party. 48 people were adults. Express 48 as a percentage of 128.

..............................% **(2 marks)**

Target grade 2

6 The table gives information about boys who chose French or German and girls who chose French or German at school.

	Number of boys	Number of girls
French	84	56
German	54	126

(a) Work out the percentage of girls who chose French.

> Start by finding the total number of girls.

..............................% **(3 marks)**

(b) 30% of the students who chose French passed their exam.
60% of the students who chose German passed their exam.
Show that 47% of all students passed their exam.

(4 marks)

55

Fractions, decimals and percentages

1 Write these percentages as fractions in their simplest form:

Per cent means out of one hundred.

Simplify your fraction if possible.

(a) 24%

$$24\% = \frac{24}{100} = \frac{\ldots}{\ldots}$$

(1 mark)

(b) 64%

(c) 72%

.... (1 mark)   .... (1 mark)

2 Write the following numbers in order, starting with the smallest:

(a) 0.62 $\frac{3}{10}$ 61%

(b) 33% 0.32 $\frac{7}{20}$

(c) 0.38 37% $\frac{2}{5}$

.... (2 marks)   .... (2 marks)   .... (2 marks)

3 John earns £2300 per month. He spends 15% of his salary on rent and $\frac{3}{5}$ of his salary on bills. Work out how much John has left after he has paid his rent and bills.

Problem solved! Plan your strategy before you start. You'll save time if you convert $\frac{3}{5}$ into a percentage.

$\frac{3}{5}$ = ... %

15% + ... % = ... %

100% − .... % = ... %

.... % of £2300 = £............

(3 marks)

4 There are 120 students in Year 11.
$\frac{9}{20}$ of the students travel to school by bike.
15% of the students travel to school by car.
The rest of the students walk to school. How many students walk to school?

.... (3 marks)

5 Amy earns £2100 each month and saves 30% of this.
Bhavna earns £1800 and saves $\frac{1}{3}$ of this.
Who saves the most money each month?

Don't just write down a name. You have to show your working and then write a conclusion.

.... (3 marks)

56

Had a go ☐   Nearly there ☐   Nailed it! ☐

RATIO & PROPORTION

Percentage change 1

1 (a) Decrease 78 by 4%. Start by finding 4% of 78.   (b) Increase 126 by 4%.

$\frac{4}{100}$ × .... = ...

78 − .... = ... **(1 mark)**   .... **(1 mark)**

(c) Decrease 96 by 12%.   (d) Increase 242 by 14%.

.... **(1 mark)**   .... **(1 mark)**

2 A shop sells mobile phones. The shop sells a mobile phone for £135.
A discount of 6% is given. Work out the price of the mobile phone after the discount.

£........................... **(2 marks)**

3 Find the percentage change of the following price changes.

	Original price (£)	New price (£)	Percentage change
(a)	640	512	$\frac{640 - ....}{640}$ × 100 = ...%
(b)	160	208	...............%
(c)	1560	2106	...............%
(d)	2750	2475	...............%

(12 marks)

4 Noah and Chloe are collecting reward points in an online video game.

Examiners' report: Use your calculator, but remember to still **write down** your working.

(a) Noah collected 3200 points last month and 4315 points this month.
Work out the percentage increase in the number of points he collected.

...............% **(3 marks)**

(b) Chloe collected 5100 points last month and 3672 points this month.
Work out the percentage decrease in the number of points she collected.

...............% **(3 marks)**

5 Niamh and Owen received the same percentage pay rise in 2015.
In 2014 Niamh earned £24 500 per year. In 2015 she received a pay rise to earn £25 970.
In 2014 Owen earned £22 000.
Work out Owen's salary in 2015.

£........................... **(4 marks)**

RATIO & PROPORTION

Had a go ☐   Nearly there ☐   Nailed it! ☐

Percentage change 2

6 Aaron is comparing the cost of flights from two airlines. Both airlines charge a credit card charge and a booking fee.

Tricky-jet
Credit card charge: 3%
Booking fee: £5

Kelly-air
Credit card charge: 5%
Booking fee: £2

A ticket is advertised as costing £90 from both airlines. Work out which airline is cheaper after the additional charges are applied.

Tricky-jet

$\dfrac{3}{\ldots\ldots} \times \ldots\ldots = \ldots\ldots$

£........... + £........... + £........... = £...........

Kelly-air

$\dfrac{5}{\ldots\ldots} \times \ldots\ldots = \ldots\ldots$

£........... + £........... + £........... = £...........

.... is cheaper.

(4 marks)

7 Zac wants to buy some concrete posts. He finds two companies on the internet.

Postland
10 posts for £10.50 each and receive a 15% discount

C & R
5 posts for £37.75 plus 20% VAT

Zac needs to buy 10 concrete posts and wants the cheapest option. Which shop should Zac buy the concrete posts from?

Examiners' report Show all your working and **write down** which shop Zac should choose. If you just circle one of them, you might not get the mark.

.... **(4 marks)**

8 Sandeep wants to buy a pair of trainers. He finds that two online shops sell the trainers he wants.

Footworld
£42.50 for a pair
Online discount of 22%

Sportish
£30.90 for a pair plus
VAT at 20%

Sandeep wants to pay the lowest price. Which shop should Sandeep buy his trainers from?

.... **(4 marks)**

Had a go ☐   Nearly there ☐   Nailed it! ☐

RATIO & PROPORTION

Ratio 1

1. Write the following ratios in their simplest form:

   (a) 45 : 30

   ÷ .... ÷ ....

   .... : ....

   (2 marks)

   Divide both parts of the ratio by the same number.

   (b) 54 : 16

   .... **(2 marks)**

   (c) 56 : 64

   .... **(2 marks)**

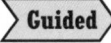

2. (a) Divide £50 in the ratio 2 : 3

   Total parts = ... + ... = ...

   1 part = 50 ÷ .... = ...

   2 parts = ... × .... = ...

   3 parts = ... × .... = ...

   (2 marks)

   (b) Divide £750 in the ratio 2 : 5 : 8

   .... **(2 marks)**

3. Sandeep is going to make a pizza. He uses cheese, peppers and dough in the ratio 2 : 3 : 7
   He uses 56 g of dough. Work out the number of grams of cheese and the number of grams of peppers he uses.

    **Problem solved!** 7 parts of the ratio represent 56 g. Work out how much one part of the ratio represents.

   cheese .... g

   pepper .... g **(3 marks)**

4. Anjali, Paul and Faye are travelling in a car from Wolverhampton to London.
   They share the driving so that the distances driven are in the ratio 3 : 4 : 5
   Anjali drives 36 miles.
   Calculate the distances Paul and Faye each drive.

   Paul .... miles

   Faye .... miles **(3 marks)**

5. Amish, Benji and Cary save some money in the ratio 3 : 4 : 9
   Cary saved £120 more than Benji.

   (a) Show that Amish saved £72.

   (b) Show that the total amount of money saved was £384.

   (2 marks) **(2 marks)**

59

RATIO & PROPORTION

Had a go ☐   Nearly there ☐   Nailed it! ☐

Ratio 2

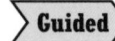

6 Solder is made from lead and tin. The ratio of the mass of lead to the mass of tin is 2 : 3

(a) Kyle made 70 g of solder. Work out the mass of the lead used.

Total parts = ... + ... = ...

1 part = 70 ÷ .... = ... g

2 parts = 2 × .... = ... g

(2 marks)

(b) He then uses 16 g of lead to make some more solder. Work out the mass of solder he made.

.... g **(2 marks)**

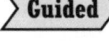

7 Gabby and Harry shared some money based on their ages. The ratio of Gabby's age to Harry's age is 3 : 8. Harry received £2000 more than Gabby. How much money did they share?

> **Problem solved!** Harry received £2000 more than Gabby and 8 − 3 = 5 so five parts of the ratio represent £2000.

8 − 3 = 5 parts

5 parts = £2000

1 part = £2000 ÷ 5 = ...

11 parts = 11 × .... = ...

In total they shared £....................

(3 marks)

8 Asha uses an old recipe to make some cakes.
The ratio of the weights of flour, margarine and sugar needed for the recipe is 5 : 4 : 3
Asha has the following amounts of each ingredient.

> 1825 g of flour
> 700 g of margarine
> 250 g of sugar

Each cake needs 48 g of the combined ingredients. Show that the maximum number of cakes she can make is 20.

(3 marks)

Had a go ☐   Nearly there ☐   Nailed it! ☐

RATIO & PROPORTION

Metric units

Target grade 2

Guided

1 Change

(a) 45 mm to cm

45 mm ÷ ....

= .... cm   **(1 mark)**

(b) 72 cm to mm

.... mm   **(1 mark)**

(c) 3.5 km to m   *Kilo means one thousand.*   (d) 5.3 kg to g

.... m   **(1 mark)**   .... g   **(1 mark)**

(e) 4.3 litres to ml   (f) 480 mg to g.

4.3 litres × .... = ... ml   **(1 mark)**   .... g   **(1 mark)**

Target grade 1

Guided

2 Change

(a) 15 cm to mm   (b) 28 mm to cm

.... mm   **(1 mark)**   28 mm ÷ .... = ... cm   **(1 mark)**

(c) 1800 g to kg   (d) 2800 m to km

.... kg   **(1 mark)**   .... km   **(1 mark)**

(e) 53 ml to litres   (f) 145 g to mg.

.... litres   **(1 mark)**   145 g × .... = ... mg   **(1 mark)**

Target grade 2

3 How many 125 ml cups can be filled from a bottle holding 2 litres of squash?

Convert 2 litres into ml.

.... **(2 marks)**

Target grade 2

4 How many 75 mm pieces of wood can be cut from a piece of wood of length 6 m?

.... **(2 marks)**

Target grade 2

5 A book shelf is 1 m wide.
Joe wants to place a set of books across the shelf.
He has 20 books. Each book is 5.2 cm wide.
Will he have enough space for all the books?

Don't just answer 'yes' or 'no'. You need to show your working and then write a conclusion.

.... **(3 marks)**

Reverse percentages

Had a go ☐ Nearly there ☐ Nailed it! ☐

1 In a sale all prices are reduced by 30%.
Andy buys a shirt on sale for £42.
Work out the original price of the shirt.

First work out the multiplier for a 30% decrease.

$$100\% - 30\% = \ldots\ldots\%$$

$$= \frac{\ldots\ldots}{100} = \ldots\ldots$$

£42 ÷ .... = £............

(3 marks)

2 Brinder receives a pay rise of 6%.
After the pay rise, Brinder earns a salary of £35 245.
Work out Brinder's salary before the pay rise.

First work out the multiplier for a 6% increase.

$$100\% + \ldots\ldots\% = \ldots\ldots\%$$

$$= \frac{\ldots\ldots}{100} = \ldots\ldots$$

£35 245 ÷ .... = £............

(3 marks)

3 Kam bought a new car. The car depreciates by 15% each year. After one year the car was worth £28 560. Work out the price of the car when it was new.

*Check that your answer makes sense. The original price of the car should be **greater** than £28 560.*

£..................... (3 marks)

4 Kate's weekly wage this year is £560.
This is 8% more than her weekly wage last year. Ken says, 'Your weekly wage was £515.20 last year.' Is Ken correct?
You must show your working.

Problem solved! You can do this question without using reverse percentages. Increase £515.20 by 8% and compare your answer to £560. Remember to write a conclusion.

.... (3 marks)

5 Alison and Nav invested some money in the stock market in 2014.
This table shows the value of their investments in 2015.

Who invested the most money originally?
Give reasons for your answer.

	Value in 2015	Percentage increase since original investment
Alison	£1848	12%
Nav	£1764	5%

.... (4 marks)

Had a go ☐ Nearly there ☐ Nailed it! ☐

RATIO & PROPORTION

Growth and decay

Tricky Topic

Target grade 5 **Guided**

1. Raj invests £12 000 for 4 years at 10% per annum compound interest.
   Work out the value of the investment at the end of four years.

   First work out the multiplier for a 10% increase.

   100% + 10% = ...%

   $\frac{............}{100}$ = ...

   £12 000 × (............)$^{.....}$ = £............................ **(2 marks)**

Target grade 5

2. Neil invests £5800 at a compound interest rate of 6% per annum.
   At the end of n complete years the investment has grown to £6907.89.
   Work out the value of n.

   Choose some values of n and work out the amount of investment after n years.

   n = ... **2 marks)**

Target grade 5 **Guided**

3. Chris bought a lorry that had a value of £24 000.
   Each year the value of the lorry depreciates by 15%.

   First work out the multiplier.

   (a) Work out the value of the lorry at the end of 4 years.

   100% − ....% = ...%

   $\frac{............}{100}$ = ...

   £24 000 × (............)$^{.....}$ = £............................ **(2 marks)**

   (b) Brian bought a new car for £12 000. Each year the value of the car depreciates by 12%. Work out the value of the car at the end of 5 years.

   £........................... **(2 marks)**

Target grade 5

4. Daljit invests £1500 on 1 January 2010 at a compound interest rate of r% per annum. The value, £V, of this investment after n years is given by the formula $V = 1500 \times (1.065)^n$

   (a) Write down the value of r.

   Work out what percentage would give a multiplier of 1.065.

   r = ... **(1 mark)**

   (b) Work out the value of Daljit's investment after 10 years.

   £........................... **(2 marks)**

Target grade 5

5. Terry buys a new vacuum cleaner for £350. The value of the machine depreciates by 25% each year. Terry says,

   '4 × 25% = 100%, so after four years the vacuum cleaner will have no value.'

   Explain why Terry is wrong.

   .... **(2 marks)**

RATIO & PROPORTION

Had a go ☐   Nearly there ☐   Nailed it! ☐

Speed

1 Anjali runs 400 metres in 44.7 seconds.
Work out Anjali's average speed.

Speed = distance ÷ time

Speed = ... ÷ .... = ... m/s

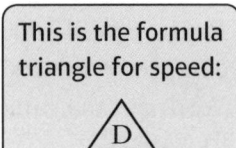
This is the formula triangle for speed:

(2 marks)

2 The distance from Manchester to Rome is 1700 km.
A plane flies from Manchester to Rome in 4 hours.
Work out the average speed of the plane.

.... km/h **(2 marks)**

3 Sandeep drives 250 km at an average speed of 75 km/h. Work out the time taken for Sandeep's journey.

.... hours .... minutes **(3 marks)**

4 Selma drives for 4 hours. Her average speed is 60 km/h.
Work out the total distance she travels.

Distance = ... × ....

Always write down the formula.

Distance = ... × .... = ... km **(2 marks)**

5 Pavan is driving in France. The legal speed limit on French motorways is 130 km/h. He travels from one junction to another in 15 minutes and he covers a distance of 35 km. Show that he has broken the speed limit.

(3 marks)

6 Jane travelled 50 km in 1 hour 15 minutes.
Karen travelled 80 km in 2 hours and 45 minutes.
Who had the lower average speed? You must show your working.

(3 marks)

7 At a school's sports day the 100 m race was won in 14.82 seconds and the 200 m race was won in 29.78 seconds. Which race was won with a faster average speed?
You must show all your working.

(3 marks)

64

Had a go ☐ Nearly there ☐ Nailed it! ☐

RATIO & PROPORTION

Density

Target grade 4 — Guided

1. What is the density of a piece of wood that has a mass of 17.5 grams and a volume of 20 cm³?

   Density = mass ÷ volume

   Density = ... ÷ ....

   = ... g/cm³   **(2 marks)**

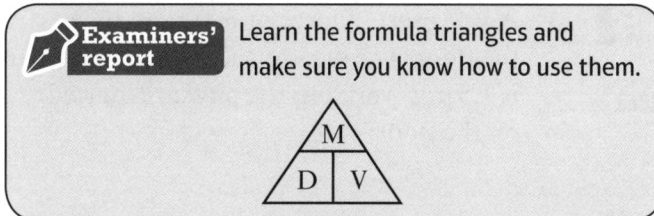

Examiners' report: Learn the formula triangles and make sure you know how to use them.

Target grade 4 — Guided

2. Len has a silver ring which has a volume of 14 cm³. The density of silver is 10.5 grams per cm³. Work out the mass of the silver ring.

   Mass = ... × ....

   Mass = ... × .... = ... g   **(2 marks)**

   Always write down the formula.

Target grade 4

3. Petrol has a density of 0.8 g/cm³. The petrol in a can has a mass of 8.3 kg. How much petrol, in cm³, does the can contain?

   Convert kg into g.

   .... cm³   **(3 marks)**

Target grade 4

4. This solid cuboid is made of plastic. The plastic has a density of 0.9 g per cm³. Work out the mass of the cuboid.

   8 cm, 10 cm, 6 cm

   .... g   **(4 marks)**

Target grade 4

5. The diagram shows a solid triangular prism. The prism is made of iron. Iron has a density of 7.87 g per cm³. Work out the mass of the prism.

   8 cm, 12 cm, 15 cm

   Remember, the volume of a prism is given by: Volume = Length × Area of cross-section.

   .... g   **(3 marks)**

Target grade 4

6. Gavin weighed some metal beads. They had a mass of 950 g. The volume of the beads was 96 cm³. Gavin worked out the density and claimed that the metal was gold.
   Use the information in the table to work out whether Gavin is correct. You must show all of your working.

Metal	Density g/cm³
Gold	19.3
Copper	8.6
Bronze	9.9

   (3 marks)

65

RATIO & PROPORTION

Had a go ☐   Nearly there ☐   Nailed it! ☐

Other compound measures

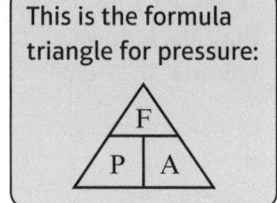

This is the formula triangle for pressure:

1 A safe exerts a force of 600 N on to the floor. The area of the base on the floor is 1.5 m². Work out the pressure exerted on the floor.

Pressure = force ÷ area

Pressure = ... ÷ .... = ... N/m²

Make sure you give units with your answer. The force is in N and the area is in m², so the units of pressure will be N/m². **(2 marks)**

2 Ray exerts a force of 900 N on to the ground. His feet have an area of 0.024 square metres each. Work out the pressure he exerts on the ground.

How many feet does Ray have?

.... N/m² **(3 marks)**

3 The pressure between a car's four tyres and the road is 400 000 N/m². The car exerts a force of 10 000 N on the road. Work out the area of contact between each tyre and the road.

Area = ... ÷ ....

Always write down the formula.

Area = ... ÷ .... = ... m² = ... m² per tyre **(2 marks)**

4 There are 140 litres of oil in a tank. When Owen opens the tank, oil flows out at a rate of 4 litres per minute. How many minutes will it take for the tank to become completely empty?

.... minutes **(2 marks)**

5 An overflow pan at a factory can be modelled as a cuboid.

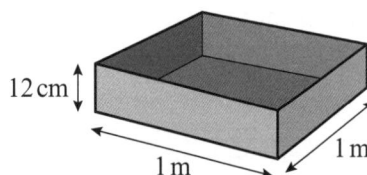

12 cm, 1 m, 1 m

Problem solved! You might be able to pick up one or two marks even on tricky questions. Here you could find the volume of the cuboid, even if you can't finish the question.

The pan is half-full of water. The water flows out of the pan at an average rate of 250 ml per second. Show that the pan will be completely empty after 4 minutes.

1 cm³ = 1 ml. Remember to convert metres to cm before calculating the volume of the cuboid.

(3 marks)

66

Had a go ☐   Nearly there ☐   Nailed it! ☐

RATIO & PROPORTION

Proportion

Target grade 2 — **Guided**

1. David buys 4 kg of apples from the supermarket for £1.60. What is the cost of 7 kg of apples?

   [Find the cost of 1 kg.]

   4 kg costs 160p

   1 kg costs .... ÷ .... = ...

   7 kg costs .... × .... = ...   **(2 marks)**

Target grade 2

2. A fabric shop sells material by the metre.
   Andy buys 3 m of material for £2.25. What is the cost of 11 m of the same material?

   £.................. **(2 marks)**

Target grade 2

3. 2 bottles of grape juice fill 8 glasses. How many glasses can be filled from 8 bottles of grape juice?

   .... glasses **(2 marks)**

Target grade 3 — **Guided**

4. 10 people take 8 days to build a wall.
   How long will it take 4 people to do the same job?

   [Work out how long it will take one person to build the wall.]

   10 people work 8 days

   1 person works .... × .... = ...

   4 people work .... ÷ .... = ... days **(2 marks)**

Target grade 3

5. A school building can be decorated by 12 people working 8 hours a day for 5 days. Mike wants to know how long it would take 10 people working 6 hours a day.

   .... **(2 marks)**

Target grade 3

6. An amount of money is divided among 8 children. Each child receives £24. If the same amount of money was divided among 12 children, how much would each child receive?

   £.................. **(2 marks)**

Target grade 3

7. A large basket of sweets costs £4.80 and holds 200 g. A medium basket of sweets costs £4.50 and holds 175 g. Which size basket is better value for money?

   [Show all your working and then write a conclusion.]

   .... **(3 marks)**

67

Proportion and graphs

1 The force, F, on a mass is directly proportional to the acceleration, a, of the mass. When $a = 25$, $F = 650$. Work out the value of F when $a = 45$

Guided

$$\frac{F}{45} = \frac{650}{25}$$

You can compare ratios to work out F.

$F = $ .... × .... = ... **(2 marks)**

2 The resistance R ohms of a wire is inversely proportional to the cross-sectional area A cm^2 of a wire. When $A = 0.1$, $R = 30$. Work out the value of R when $A = 0.4$.

Guided

$R \times 0.4 = $ .... × ....

$R = $ .... ÷ .... = ... **(2 marks)**

3 x and y are inversely proportional. Circle the equation that could describe the relationship between x and y.

$x = 2y \qquad x = 3\sqrt{y} \qquad x = \dfrac{1}{2y} \qquad x = \dfrac{y}{5}$

(1 mark)

4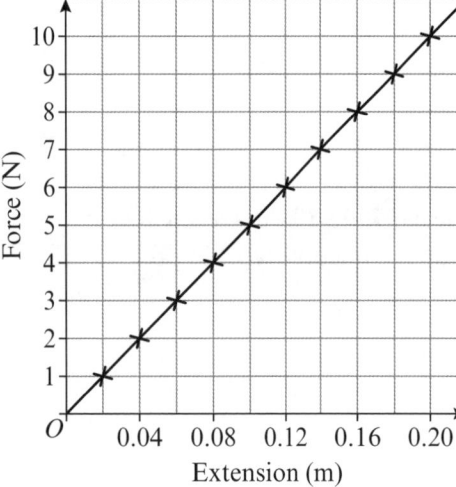

This graph shows the relationship between the extension, in m, of a spring and the force, in N.

(a) Use the graph to find the force when the extension is 0.1 m.

.... **(1 mark)**

(b) What evidence is there from the graph to show that force is directly proportional to extension?

.... **(2 marks)**

5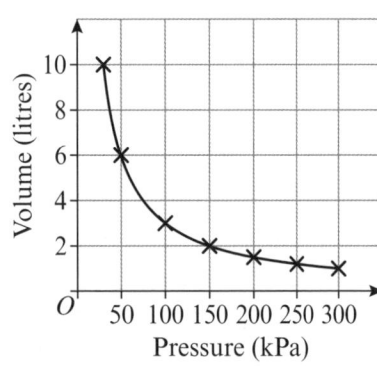

This graph shows the relationship between pressure, in kPa, and volume, in litres.

(a) Use the graph to find the volume when the pressure is 150 kPa.

.... **(1 mark)**

(b) What evidence is there from the graph to show that volume is inversely proportional to pressure?

.... **(2 marks)**

Had a go ☐   Nearly there ☐   Nailed it! ☐

RATIO & PROPORTION

Problem-solving practice 1

1 Julie got 41 out of 50 marks in a mathematics test.
She got 50 out of 60 marks in a statistics test.
In which test did Julie get the higher percentage mark?
You must show all your working.

(3 marks)

2 There are 500 students at Pennhouse School.
122 students were absent from school on Monday.
Rachael says that more than 25% of the students
were absent on Monday.
Is she correct? Explain your answer.

.... (3 marks)

3 Karen wants to buy a game for her new games console.
She finds that two online shops sell the game she wants.

Nile	T-bay
Game costs £35.50	Game costs £30.90 + VAT
Online discount of 16%	VAT at 20%
Delivery charge of £2.75	No delivery charge

Karen wants to pay the lowest price.
Which shop should Karen buy her game from? You must show all your working.

(4 marks)

4 Avtar has a full 900 ml bottle of patio sealant. He is going to mix some of the patio sealant with water. Here is the information on the label of the bottle.

Patio sealer (900 ml)
Mix $\frac{1}{5}$ of the patio sealant with 5400 ml of water

Avtar is going to use 900 ml of water. How many millilitres of patio sealant should Avtar use? You must show your working.

(4 marks)

RATIO & PROPORTION

Had a go ☐ Nearly there ☐ Nailed it! ☐

Problem-solving practice 2

5 Taran employs 8 people to plaster a building in 6 days.
He realises that he needs to plaster the building in just 4 days.
Taran says that he needs 3 more people working at the same rate to plaster the building in 4 days.
Is he correct? You must show your working.

(2 marks)

6 Brett is going to buy some bird food. Bird food is sold in 200 g boxes costing £2.50 and in 1000 g boxes costing £10.50. Which box of bird food gives the better value for money? You must show your working.

(3 marks)

7 This is a list of ingredients for making an apple and almond crumble for 4 people.
Rachel has the following ingredients:

Ingredients for 4 people	Rachel's ingredients
80 g flour	1.2 kg flour
60 g almonds	200 g almonds
90 g brown sugar	500 g brown sugar
60 g butter	250 g butter
4 apples	16 apples

She wants to make an apple crumble for 15 people.
Does she have enough ingredients?
Show all of your working.

(3 marks)

8 Kim wants to save a deposit for a house. His target is to save £17 500 in 4 years.
He invests £14 000 in an ISA for 4 years at 6% per annum compound interest.
Does he have enough money for his deposit? You must show your working.

(3 marks)

Had a go ☐ Nearly there ☐ Nailed it! ☐

GEOMETRY & MEASURES

Symmetry

Target grade 1

Guided

1 The following shapes have lines of symmetry. Draw the lines of symmetry as indicated.

(a) exactly **one** line of symmetry

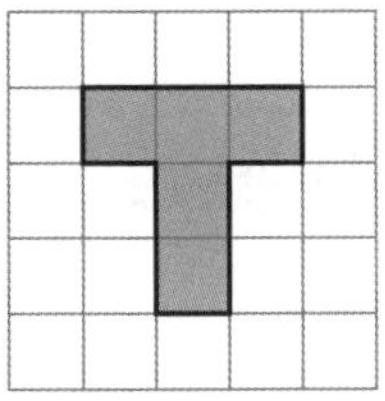

(b) exactly **two** lines of symmetry

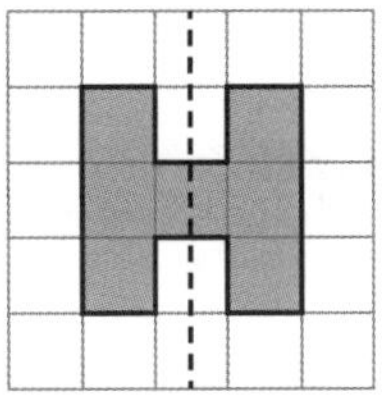

(1 mark)   **(1 mark)**

Target grade 1

2 Here are four shapes.

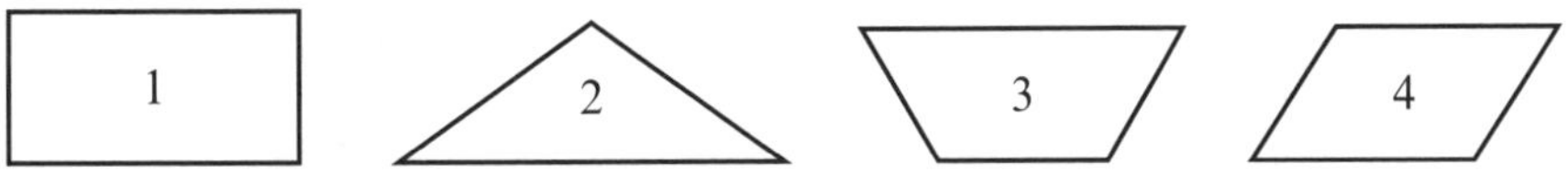

   1   2   3   4

Write down the number of a shape that has

(a) **no** lines of symmetry

(b) exactly **one** line of symmetry

(c) exactly **two** lines of symmetry.

.... **(1 mark)**   .... **(1 mark)**   .... **(1 mark)**

Target grade 2

3 On the diagram below, shade one square so that the shape has

(a) exactly **one** line of symmetry

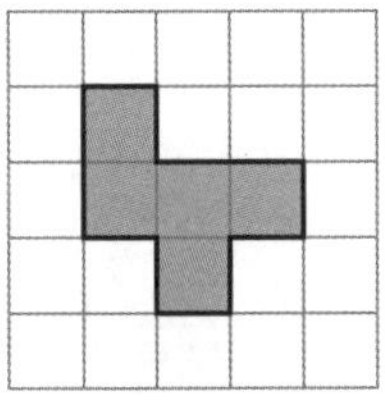

(b) rotational symmetry of order 2

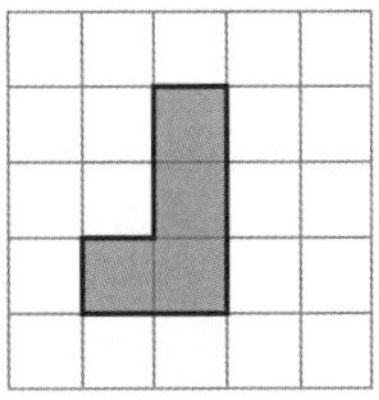

(1 mark)   **(1 mark)**

(c) exactly **one** line of symmetry

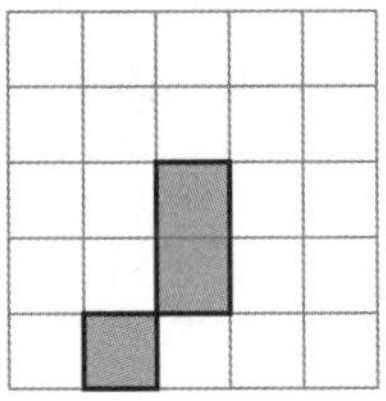

(d) rotational symmetry of order 2.

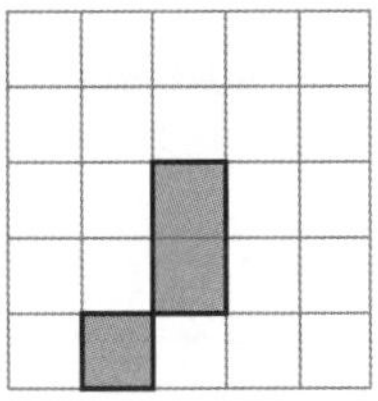

(1 mark)   **(1 mark)**

Target grade 2

4 Here is a regular hexagon.

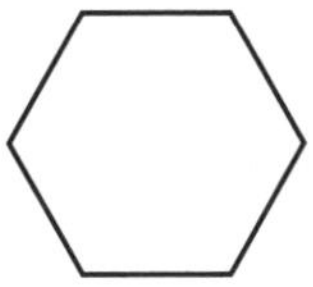

Problem solved! Remember you can use tracing paper in your exam. This can help you check rotational symmetry.

(a) What is the order of rotational symmetry of the hexagon?

.... **(1 mark)**

(b) Draw a line of symmetry on the hexagon.   **(1 mark)**

Quadrilaterals

1 Write down the mathematical name of each of these quadrilaterals.

(a) 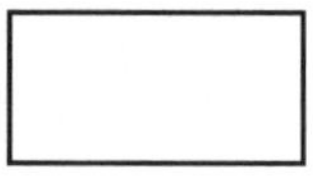 Rectangle (1 mark)

(b) 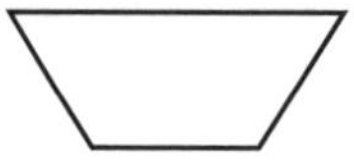 .... (1 mark)

(c) 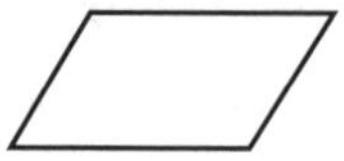 .... (1 mark)

(d) 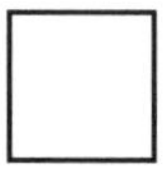 .... (1 mark)

(e) 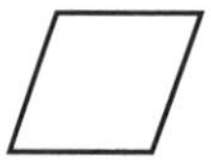 .... (1 mark)

(f) 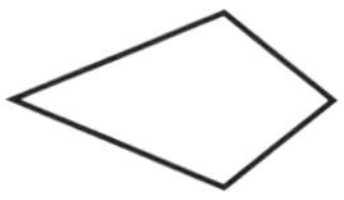 .... (1 mark)

2 Draw a quadrilateral with the following properties:

(a) one pair of parallel sides

> **Examiners' report**: Use arrows to show parallel lines, and dashes to show lines of equal length.

(1 mark)

(b) two pairs of parallel sides, all sides equal and diagonals cross at right angles

(1 mark)

(c) two pairs of adjacent sides equal, one pair of opposite angles equal and diagonals cross at 90°

(1 mark)

(d) two pairs of parallel sides, opposite sides equal and opposite angles equal

(1 mark)

(e) all angles 90°, opposite sides equal and parallel and diagonals equal

(1 mark)

(f) all angles 90°, all sides equal, opposite sides parallel, diagonals equal and cross at right angles.

(1 mark)

Had a go ☐ Nearly there ☐ Nailed it! ☐

GEOMETRY & MEASURES

Angles 1

1 Draw lines from the name of the angle to the correct diagram.

acute   obtuse   reflex

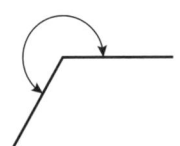 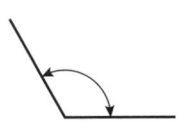 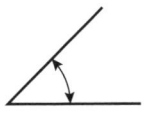

(3 marks)

2 (a) 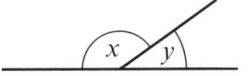   (b)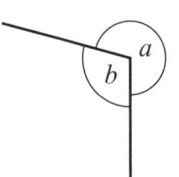

Use one of the words given in question 1 above.

(i) What type of angle is x?

....
(1 mark)

(ii) Give a reason for your answer.

....
(1 mark)

(i) What type of angle is a?

....
(1 mark)

(ii) Give a reason for your answer.

....
(1 mark)

3 (a)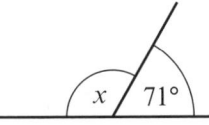

(i) Work out the size of the angle marked x.

.... − 71° = ...° **(1 mark)**

(ii) Give a reason for your answer.

Angles on a straight line add up to ....° **(1 mark)**

(b)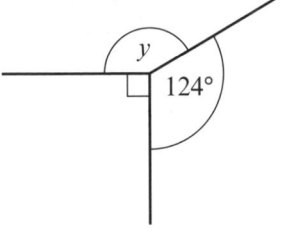

A right angle is 90°.

(i) Work out the size of the angle marked y.

..................° **(1 mark)**

(ii) Give a reason for your answer.

Angles around a point add up to ....° **(1 mark)**

73

GEOMETRY & MEASURES

Had a go ☐ Nearly there ☐ Nailed it! ☐

Angles 2

4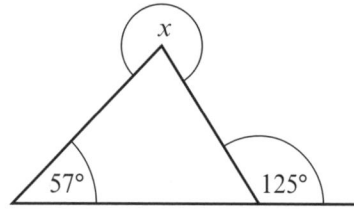

Work out the size of the angle marked *x*.

Angles on a straight line add up to 180°, angles around a point add up to 360° and angles in a triangle add up to 180°.

.... − 125° = ...°

180° − 57° − ....° = ...°

x = 360° − ....° = ...°

x = ...

(3 marks)

5 The diagram shows a five-sided shape. All the sides are equal in length.

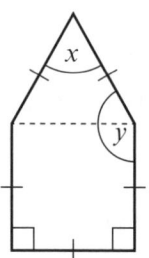

(a) (i) Work out the size of the angle marked *x*.

.... ÷ ....

= ....°

(1 mark)

(ii) Give a reason for your answer.

The triangle at the top of the shape is an

.... triangle. **(1 mark)**

(b) Work out the size of the angle marked *y*.

..................................° **(2 marks)**

6 Work out the size of each marked angle.
Give a reason for your answers.

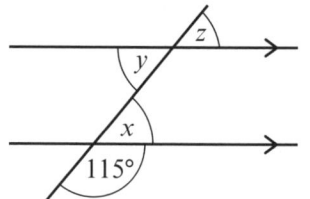

Angles *x* and *y* are alternate angles between parallel lines.

(a) *x* = ...°

Reason .... **(1 mark)**

(b) *y* = ...°

Reason .... **(1 mark)**

(c) *z* = ...°

Reason .... **(1 mark)**

Had a go ☐ Nearly there ☐ Nailed it! ☐

GEOMETRY & MEASURES

Solving angle problems

Target grade 2

Guided

1 (a)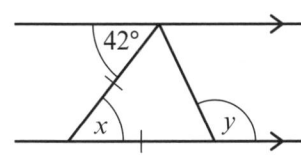

(i) Work out the size of the angle marked x.

$x = $ ....° **(1 mark)**

(ii) Give a reason for your answer.

.... angles are equal **(1 mark)**

(iii) Work out the size of the angle marked y. Give reasons for each step of your working.

$180 - x = $ ....

.... $\div 2 = $ ....

$y = 180 - $ .... = ... **(2 marks)**

(b)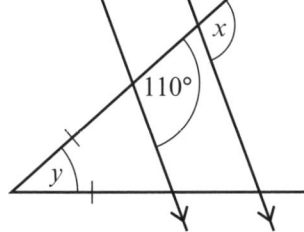

(i) Work out the size of the angle marked x.

$x = $ ....° **(1 mark)**

(ii) Give a reason for your answer.

.... **(1 mark)**

(iii) Work out the size of the angle marked y.

$y = $ ....° **(1 mark)**

(iv) Give a reason for your answer.

.... **(1 mark)**

Target grade 3

Guided

2 The diagram shows three straight lines. Work out the value of x.

Corresponding angles are equal, and opposite angles are equal.

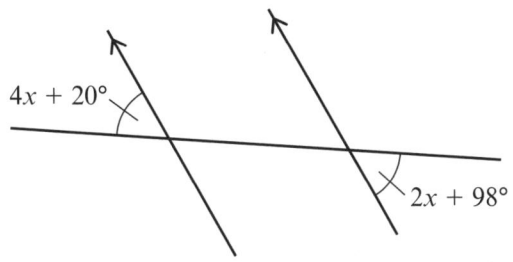

$4x + 20 = $ ....

Examiners' report: Trial and improvement is unlikely to work. You need to form an equation and solve it. You can gain a method mark just by forming your equation.

$x = $ .... **(3 marks)**

75

Angles in polygons

Had a go ☐ **Nearly there** ☐ **Nailed it!** ☐

1. The diagrams show regular polygons.
   Work out the size of an exterior angle for each regular polygon.

   (a)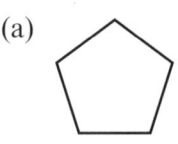

   Exterior angle = 360 ÷ number of sides

   Exterior angle = 360 ÷ .... = ...° **(2 marks)**

   (b)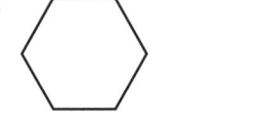

   ....° **(2 marks)**

   (c)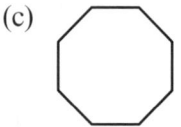

   ....° **(2 marks)**

2. The interior angle of a regular polygon is 140°.

   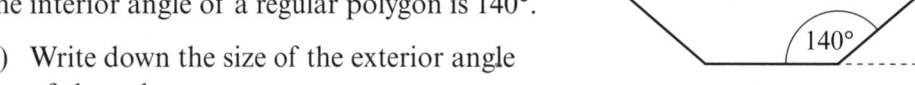

   (a) Write down the size of the exterior angle of the polygon.

   Exterior angle = 180 − .... = ...° **(1 mark)**

   (b) Work out the number of sides of the polygon.

   Number of sides = 360 ÷ .... = ... **(2 marks)**

3. Each diagram shows part of a regular polygon.
   The size of one interior angle is given.
   Work out the number of sides in each polygon.

   Problem solved! Work out the size of one exterior angle, then divide 360° by this.

   (a)  150°

   .... **(3 marks)**

   (b) 144°

   .... **(3 marks)**

   (c)  162°

   .... **(3 marks)**

4. The diagram shows part of a regular octagon.
   Work out the size of the angle marked x.

   ....° **(3 marks)**

5. The diagram shows three sides of a regular hexagon. Show that $x = 30°$.

   (3 marks)

Had a go ☐ Nearly there ☐ Nailed it! ☐

GEOMETRY & MEASURES

Time and timetables

Target grade 1
Guided

1. Write down the following times using the 24-hour clock:

   (a) 3.15 pm (b) 2.25 am (c) 11.48 pm.

   ....:15 **(1 mark)** ....:25 **(1 mark)** .... **(1 mark)**

Target grade 1

2. The following times are given using the 24-hour clock. Write the times using am or pm.

   (a) 04:25 (b) 12:10 (c) 20:32

   .... **(1 mark)** .... **(1 mark)** .... **(1 mark)**

Target grade 1
Guided

3. A train sets off from Coventry at 08:25 and arrives at London Euston at 09:17. How long does the journey take?

   08.25 to 09.00 to 09.17

   .... mins + ... mins = ... mins

   Break the journey up.

   .... minutes **(2 marks)**

Target grade 1

4. A cyclist sets off at 10.25 am and cycles for one and three quarter hours. He then rests for 35 minutes and returns home by a different route which takes 2 hours and 45 minutes. What time does he arrive back home?

   Give your answer using the 24-hour clock.

   .... **(2 marks)**

Target grade 1

5. Here is part of a timetable.

Train	A	B	C	D	E
Wolverhampton	06:45	07:05	07:25	07:45	08:10
London	08:35	08:56	09:15	09:34	10:15

   (a) Which train took more than two hours to go from Wolverhampton to London?

   .... **(1 mark)**

   (b) Work out the number of minutes taken by train B to go from Wolverhampton to London.

   .... minutes **(2 marks)**

   (c) Aaron has a meeting in London. He needs to arrive in London before 10:00. Write down the time of the latest train he can catch.

   .... **(1 mark)**

   (d) Train B arrives 36 minutes late in London. What time does this train arrive in London?

   .... **(1 mark)**

GEOMETRY & MEASURES

Had a go ☐   Nearly there ☐   Nailed it! ☐

Reading scales

Target grade 1

Guided

1   Write down the number marked with an arrow.   Look at the scales carefully.

(a)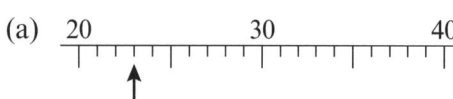

2.......... **(1 mark)**

(b)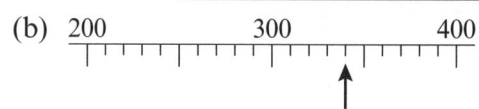

3.......... **(1 mark)**

(c)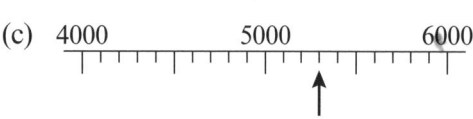

.... **(1 mark)**

(d)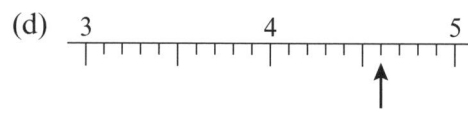

.... **(1 mark)**

Target grade 1

2   Find the following numbers on the number line. Mark them with an arrow (↑).

(a) 33

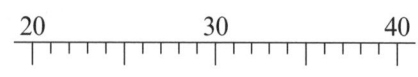

(1 mark)

(b) 280

(1 mark)

(c) 4700

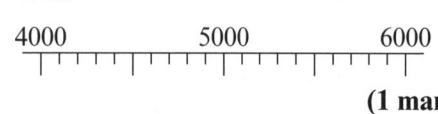

(1 mark)

(d) 3.8

(1 mark)

Target grade 1

3   (a) Write down the speed marked with an arrow.

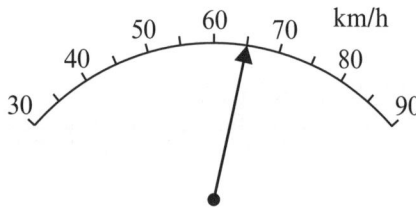

.... km/h   **(1 mark)**

(b) Mark 42 km/h with an arrow (↑).

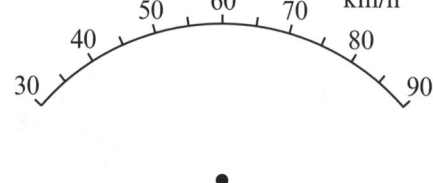

(1 mark)

Target grade 1

4   The diagram below shows 6 identical cubes and 4 identical triangles.

Problem solved! Use the first scale to work out the weight of one square.

Work out the weight, in kg, of one triangle.

.... kg   **(3 marks)**

78

Had a go ☐ Nearly there ☐ Nailed it! ☐

GEOMETRY & MEASURES

Perimeter and area

Target grade 1

Guided

1 A shape has been drawn on a grid of centimetre squares.

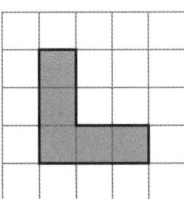

Count the number of shaded squares to find the area.

Count around the shape to find the perimeter.

(a) Find the area of the shaded shape.

Area = ... cm^2 **(1 mark)**

(b) Find the perimeter of the shaded shape.

Perimeter = ... cm   **(1 mark)**

Target grade 1

Guided

2 Two shapes have been drawn on a grid of centimetre squares.

Count 1 cm^2 for every whole square and $\frac{1}{2}$ cm^2 for every part square.

(a) 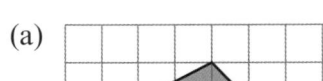   Work out the area of the shape.

Area of whole squares = ... cm^2

Area of part squares = ... cm^2

Total area = ... + ... = ... cm^2 **(1 mark)**

(b) 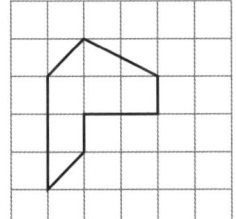   Work out the area of the shape.

.... cm^2 **(1 mark)**

Target grade 2

3 Work out the perimeter of

(a) this rectangle

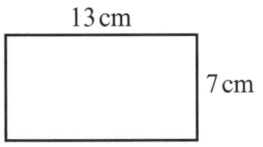

13 cm, 7 cm

.... cm   **(1 mark)**

(b) this trapezium

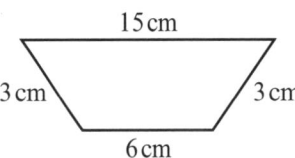

15 cm, 3 cm, 3 cm, 6 cm

.... cm   **(1 mark)**

(c) this regular hexagon

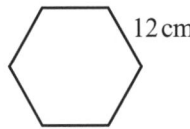

12 cm

.... cm   **(1 mark)**

(d) this parallelogram.

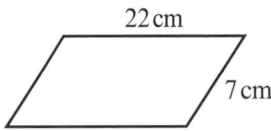

22 cm, 7 cm

.... cm   **(1 mark)**

79

GEOMETRY & MEASURES

Had a go ☐ Nearly there ☐ Nailed it! ☐

Area formulae

1 Find the areas of the following shapes.

(a)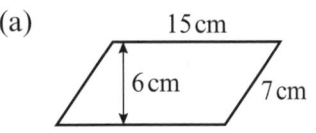

Area = ... × ....

Area = ... cm² **(2 marks)**

(b)

Area = ½(.............. + ...) × ....

Area = ... cm² **(2 marks)**

(c)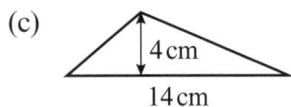

Area = ½ × .... × ....

Area = ... cm² **(2 marks)**

(d)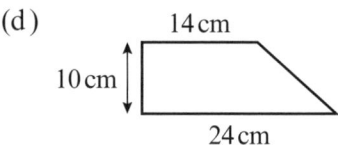

Area = ½(.............. + ...) × ....

Area = ... cm² **(2 marks)**

2 (a) The length of a rectangle is twice its width. The length is 12 cm.
Work out the area of the rectangle.

> Draw a diagram and then write the length and width on it.

.... cm² **(2 marks)**

(b) The base of a right-angled triangle is four times its vertical height.
The vertical height is 14 cm. Work out the area of the right-angled triangle.

.... cm² **(2 marks)**

3 The area of the parallelogram is three times the area of the trapezium.

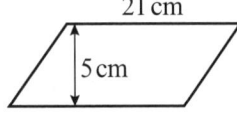 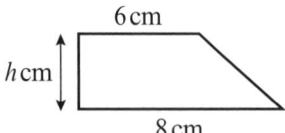

Find the height, h cm, of the trapezium. You must show all your working.

.... cm²

(3 marks)

Had a go ☐ Nearly there ☐ Nailed it! ☐

GEOMETRY & MEASURES

Solving area problems

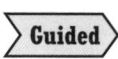

1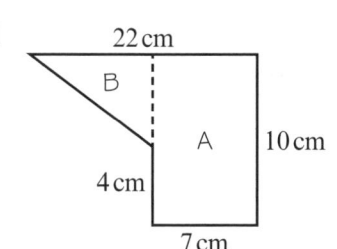

The diagram shows a compound shape.
Work out the area of the shape.

> Divide the shape into a rectangle and a triangle. Label the shapes A and B.

Area A = ... × .... = ... cm²

Area B = $\frac{1}{2}$ × .... × .... = ... cm²

Total area = Area A + Area B = ... + ...

= .... cm² **(3 marks)**

2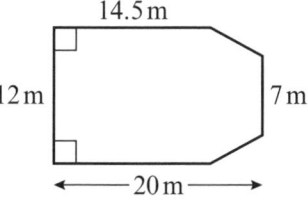

> **Examiners' report** Read the question carefully to work out whether you need to find the perimeter or the area. The units are **square metres** so this is an area question.

The diagram shows the plan of a car park.
The council wants to sell the car park.
The council wants at least £27 per square metre.
A local developer offers £6000. Will the council accept this offer? You must show your working.

.... **(5 marks)**

3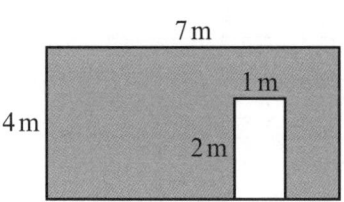

The diagram shows a wall with a door in it.
Amanda wants to paint the wall.
She buys five tins of paint.
Each tin covers 1.5 m² of wall.
Does she have enough tins to paint the wall?
Give a reason for your answer.

.... **(3 marks)**

81

GEOMETRY & MEASURES

Had a go ☐ Nearly there ☐ Nailed it! ☐

3-D shapes

1 Write down the names of these shapes.

(a)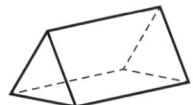

Cube .... **(1 mark)**

(b)

Cub................ **(1 mark)**

(c)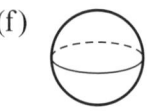

.... **(1 mark)**

(d)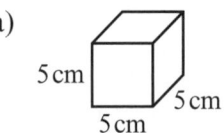

.... **(1 mark)**

(e)

.... **(1 mark)**

(f)

Sp.............. **(1 mark)**

2 Find the surface area of the following shapes.

(a) 5cm, 5cm, 5cm

Area of 1 face = ... × .... = ... cm²

Area of 6 faces = 6 × .... = ... cm²   **(2 marks)**

(b) 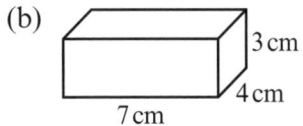 7cm, 4cm, 3cm

.... cm² **(2 marks)**

(c) 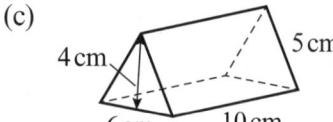 4cm, 5cm, 6cm, 10cm

.... cm² **(2 marks)**

3 Complete the table.

	3-D shape	Number of faces	Number of edges	Number of vertices
(a)	Cube			
(b)	Cuboid			
(c)	Triangular prism			
(d)	Tetrahedron			

(4 marks)

4 The diagram shows a cube.  4cm

(a) Work out the total surface area of the cube.

.... cm² **(2 marks)**

(b) Sara wants to paint all 6 faces of 40 of these cubes.
Each tin of paint covers an area of 350 cm². She buys 10 tins.
Does she buy enough tins?
Give a reason for your answer.

.... **(3 marks)**

Had a go ☐   Nearly there ☐   Nailed it! ☐   **GEOMETRY & MEASURES**

Volumes of cuboids

1 Find the volumes of the following cuboids.

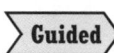

(a)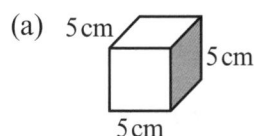

Volume = ... × .... × ....

Volume = ... cm³   **(2 marks)**

(b) 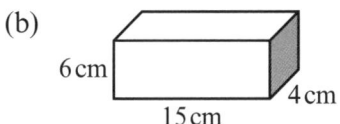   (c)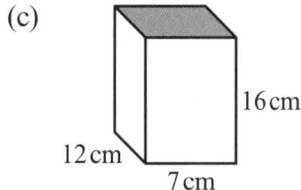

.... cm³ **(2 marks)**   .... cm³ **(2 marks)**

2 A cuboid has a volume of 504 cm³, a length of 12 cm and a width of 7 cm.
   Work out the height of the cuboid.

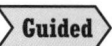

Volume = length × width × height

.... = ... × .... × height

height = ... ÷ .... = ... cm   **(2 marks)**

3 A box measures 175 cm × 125 cm × 100 cm. The box is to be completely filled
   with cubes. Each cube measures 25 cm × 25 cm × 25 cm.
   Work out the number of cubes that can completely fill the box.

   > Work out how many cubes can fit along each dimension of the box.

.... cubes **(3 marks)**

4 The diagrams show a rectangular tray and a carton.

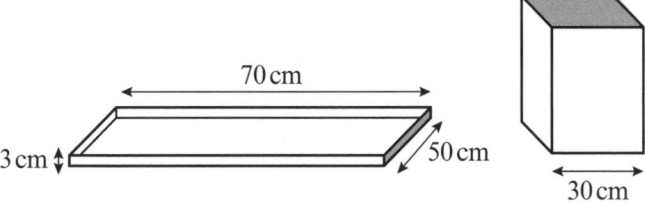

> **Problem solved!** Find the volume of the tray. You need to find the height of a cuboid with a 30 cm × 30 cm base that has the same volume as the tray.

The rectangular tray has length 70 cm, width 50 cm and depth 3 cm.
The carton is a cuboid with a square base of side 30 cm.
The tray is full of water. The water is poured into the empty carton.
Work out the depth, in cm, of the water in the carton.

.... cm **(3 marks)**

83

GEOMETRY & MEASURES

Had a go ☐ Nearly there ☐ Nailed it! ☐

Prisms

Target grade 4

Guided

1. Find the volumes of the following prisms.

   (a)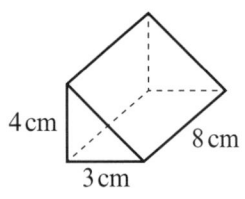

   > Volume of a prism = area of cross-section × length
   > You need to learn this formula for your exam.

   Volume = ($\frac{1}{2}$ × .... × ....) × ....

   Volume = ... cm³   **(2 marks)**

   (b)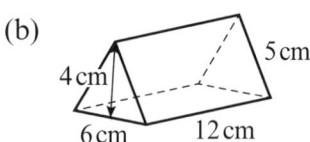

   .... cm³   **(2 marks)**

   (c)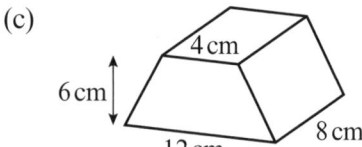

   .... cm³   **(2 marks)**

Target grade 4

Guided

2. Find the surface areas of the following prisms.

   (a)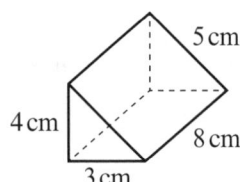

   Surface area = 2 ($\frac{1}{2}$ × .... × ....)

   + (............... × ....) + (............... × ....)

   + (............... × ....)

   Surface area = ... cm²   **(3 marks)**

   (b)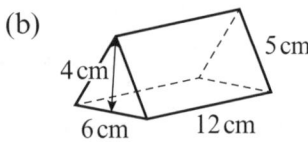

   .... cm²   **(2 marks)**

   (c)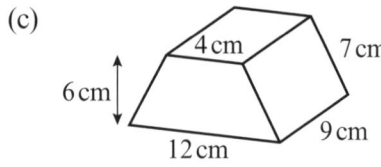

   .... cm²   **(2 marks)**

Target grade 4

3. The diagram shows a prism.
   The area of the cross-section of the prism is 22 cm².
   The length of the prism is 16 cm.
   Work out the volume of the prism.

   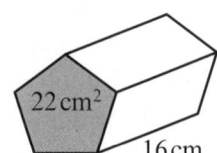

   .... cm³   **(2 marks)**

84

Had a go ☐   Nearly there ☐   Nailed it! ☐

GEOMETRY & MEASURES

Units of area and volume

Target grade 4 **Guided**

1 Convert

(a) $6\,m^2$ into cm^2

$6\,m^2 = 6 \times \rule{2cm}{0.4pt} = \rule{2cm}{0.4pt}\,cm^2$ **(2 marks)**

(b) $15\,cm^2$ into mm^2 (c) $4\,km^2$ into m^2 (d) $500\,000\,cm^2$ into m^2

……………… mm^2 **(2 marks)**   ……………… m^2 **(2 marks)**   ……………… m^2 **(2 marks)**

(e) $60\,000\,mm^2$ into cm^2 (f) $800\,000\,m^2$ into km^2.

……………… cm^2 **(2 marks)**   ……………… km^2 **(2 marks)**

Target grade 4 **Guided**

2 Convert

(a) $22\,m^3$ into cm^3

$22\,m^3 = 22 \times \rule{2cm}{0.4pt} \times \rule{2cm}{0.4pt} \times \rule{2cm}{0.4pt}$

$= \rule{2cm}{0.4pt}\,cm^3$ **(2 marks)**

(b) $28\,cm^3$ into mm^3 (c) $3\,km^3$ into m^3 (d) $200\,000\,000\,cm^3$ into m^3

……………… mm^3 **(2 marks)**   ……………… m^3 **(2 marks)**   ……………… m^3 **(2 marks)**

(e) $50\,000\,000\,mm^3$ into cm^3 (f) $420\,000\,000\,m^3$ into km^3.

……………… cm^3 **(2 marks)**   ……………… km^3 **(2 marks)**

Target grade 4

3 Convert

(a) $200\,000\,cm^3$ into litres   (b) $8\,m^3$ into litres   (c) $12\,m^3$ into litres.

……………… litres **(2 marks)**   ……………… litres **(2 marks)**   ……………… litres **(2 marks)**

Target grade 4

4 Work out how many litres of water each tank in the shape of a cuboid can hold.

(a) 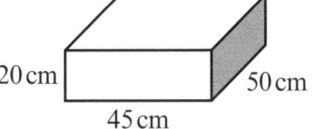   (b)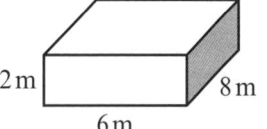

……………… litres **(2 marks)**   ……………… litres **(2 marks)**

GEOMETRY & MEASURES

Had a go ☐   Nearly there ☐   Nailed it! ☐

Translations

Target grade 3

Guided

1 Write down the following information in vector notation.

(a) 3 to the right
    4 up

$\begin{pmatrix} 3 \\ \end{pmatrix}$ **(1 mark)**

(b) 2 to the left
    3 down

$\begin{pmatrix} - \\ - \end{pmatrix}$ **(1 mark)**

(c) 5 to the left
    6 up

$\begin{pmatrix} \\ \end{pmatrix}$ **(1 mark)**

Target grade 3

2 (a) Translate shape A 3 squares right and 4 squares down. Label the image C.

(2 marks)

(b) Translate shape B 5 squares left and 7 squares up. Label the image D.

(2 marks)

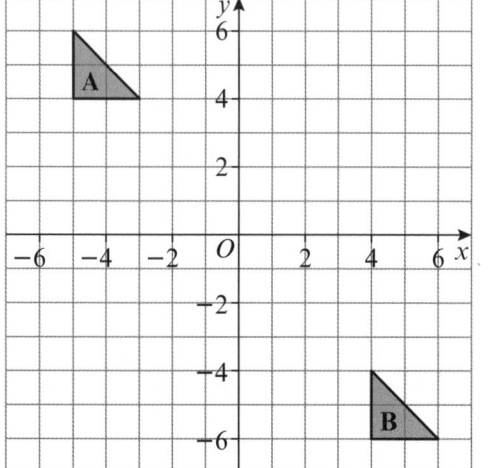

Target grade 3

3 (a) Describe fully the single transformation that will map shape G onto shape H.

....

....

....

(b) Describe fully the single transformation that will map shape M onto shape N.

....

....

(2 marks)

(2 marks)

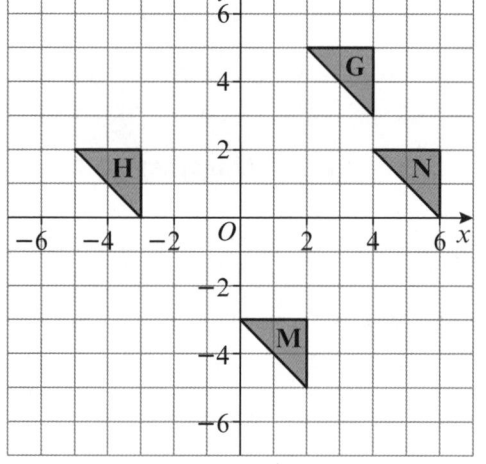

Had a go ☐   Nearly there ☐   Nailed it! ☐

GEOMETRY & MEASURES

Reflections

1 Reflect the shaded shape in the mirror line (indicated by a dashed line).

> The reflected shape should be the same distance from the mirror line as the original shape.

(a)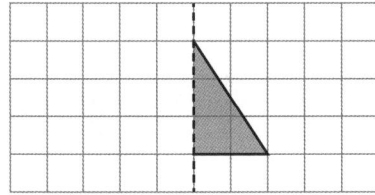

(1 mark)

(b)

(1 mark)

(c)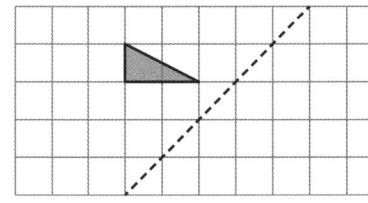

(1 mark)

2 (a) Reflect shape P in the line $y = 1$
       Label the image Q.

(2 marks)

(b) Reflect shape R in the line $x = 1$
    Label the image S.

(2 marks)

(c) Reflect shape T in the line $y = x$
    Label the image U.

(2 marks)

 Make sure you are confident drawing and naming vertical ($x = ...$) and horizontal ($y = ...$) lines.

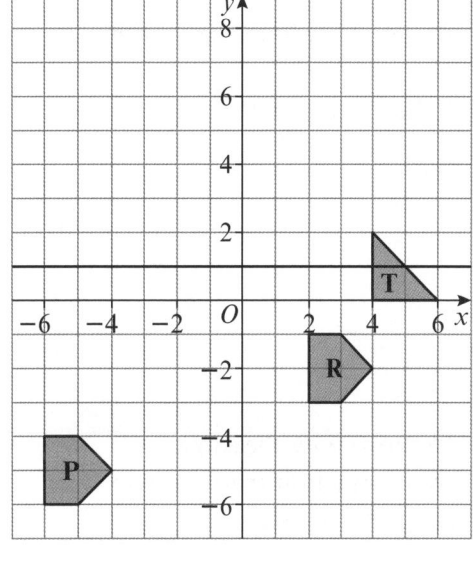

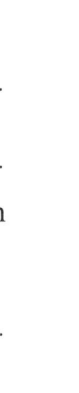

3 (a) Describe fully the single transformation that will map shape A onto image B.

....

....

(2 marks)

(b) Describe fully the single transformation that will map image B onto image C.

....

....

(3 marks)

87

GEOMETRY & MEASURES

Had a go ☐   Nearly there ☐   Nailed it! ☐

Rotations

1 (a) Rotate triangle A 180° about the point (−1, 2).
      Label the image B.   (2 marks)

   > Trace triangle A using tracing paper. Put your pencil at the point (−1, 2) and then rotate the tracing paper through 180°.

   (b) Rotate triangle A 90° anticlockwise about the point (2, 1).
       Label the image C.   (2 marks)

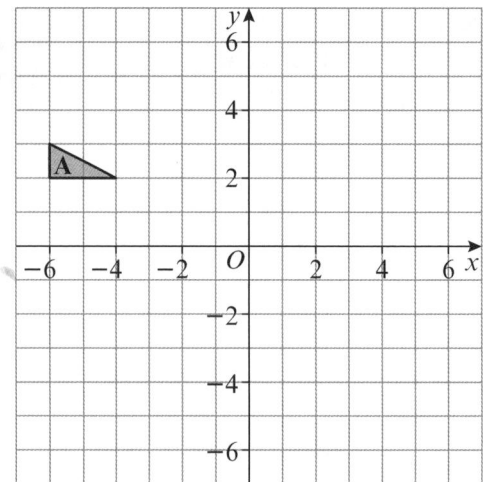

2 (a) Rotate shape P 90° clockwise about the point (0, 1).
       Label the image Q.   (2 marks)

   (b) Rotate shape P 180° about the point (−2, −1).
       Label the image R.   (2 marks)

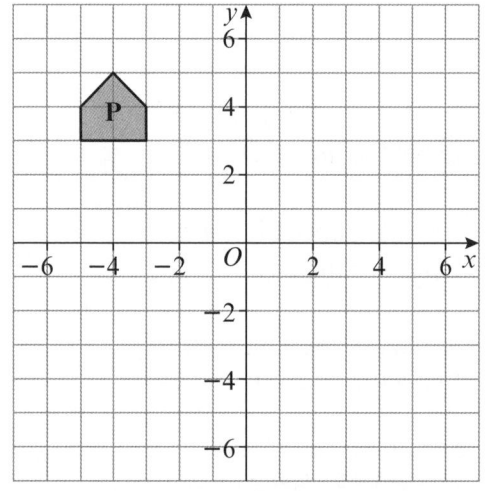

3 (a) Describe fully the single transformation that will map shape A onto shape B.

   ....

   ....

   ....   (3 marks)

   (b) Describe fully the single transformation that will map shape A onto shape C.

   ....

   ....

   ....   (3 marks)

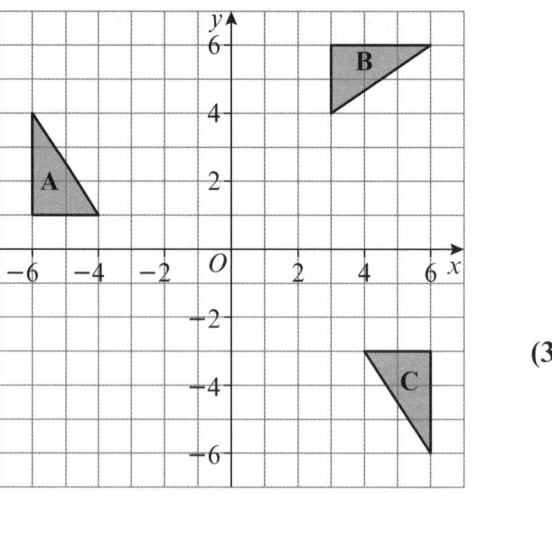

Had a go ☐   Nearly there ☐   Nailed it! ☐

GEOMETRY & MEASURES

Enlargements

Target grade 3 **Guided**

1 (a) Shape B is an enlargement of shape A. Find the scale factor of the enlargement.

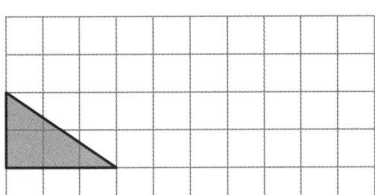

.... ÷ .... = ... **(1 mark)**

(b) Enlarge the triangle below by scale factor 2.

> No centre of enlargement is given so the enlarged shape can be placed anywhere on the grid.

(1 mark)

Target grade 3 **Guided**

2 Enlarge shape A by scale factor 3, centre (0, 0).
Label the image B.

> Draw lines from the centre of enlargement through each vertex of shape A.

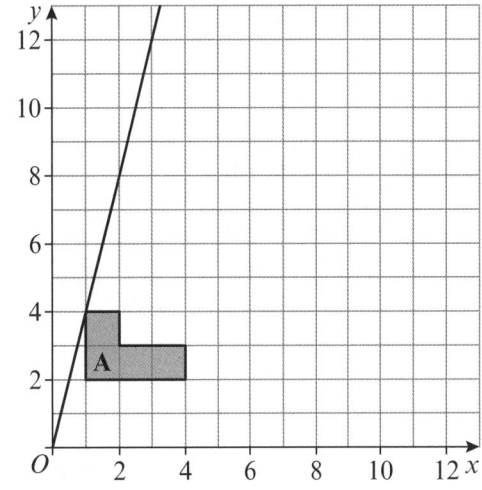

(2 marks)

Target grade 5

3 (a) Describe fully the single transformation that will map shape A onto shape B.

> The image is **smaller** than the object, so the scale factor will be a **fraction**.

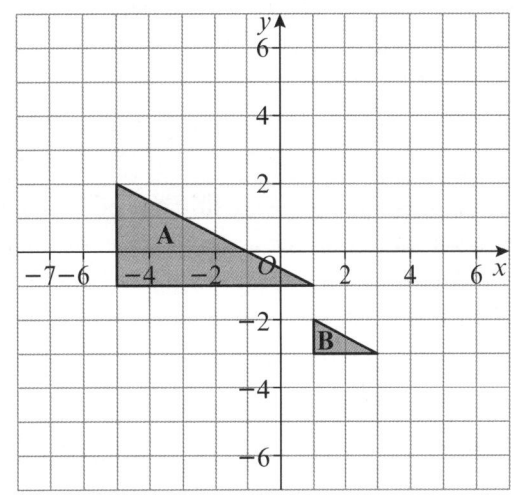

....

....

....

(3 marks)

(b) Enlarge shape A with scale factor $\frac{1}{2}$, centre (−7, −5).

(2 marks)

89

GEOMETRY & MEASURES

Had a go ☐   Nearly there ☐   Nailed it! ☐

Pythagoras' theorem

Target grade 4

Guided

1 Work out the lengths of the sides marked with letters in the following triangles.
Give your answers correct to 3 significant figures.

$\boxed{\text{short}^2 + \text{short}^2 = \text{long}^2}$

(a)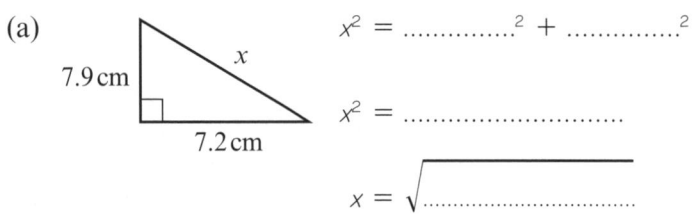

$x^2 = \ldots\ldots^2 + \ldots\ldots^2$

$x^2 = \ldots\ldots\ldots\ldots$

$x = \sqrt{\ldots\ldots\ldots\ldots}$

$x = \ldots\ldots$ cm   **(2 marks)**

(b)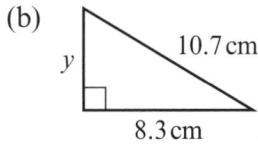

$y = \ldots\ldots$ cm   **(2 marks)**

(c)

$\ldots\ldots^2 = z^2 + \ldots\ldots^2$

$z^2 = \ldots\ldots^2 - \ldots\ldots^2$

$z = \sqrt{\ldots\ldots\ldots\ldots}$

$z = \ldots\ldots\ldots\ldots$ cm   **(2 marks)**

Target grade 4

2 One end of a rope is tied to the top of a vertical flagpole of height 12.8 m. When the rope is pulled tight, the other end is on the ground 4.2 m from the base of the flagpole. Work out the length of the rope.
Give your answer correct to the nearest cm.

Examiners' report If no diagram is given with a question, it sometimes helps to sketch your own.

.... cm   **(2 marks)**

Target grade 4

3 Cindy has a rectangular suitcase of length 95 cm and width 72 cm.
She wants to put her walking stick into her suitcase.
The length of the walking stick is 125 cm.
She thinks that the walking stick will fit into her suitcase. Is she correct?
Give a reason for your answer.

.... **(2 marks)**

Target grade 4

4 The diagram shows a small pool with a radius of 2.8 m and a height of 1.5 m.
A straight pole is 6 m long.
The pole cannot be broken.
Can the pole be totally immersed in the pool?
Give a reason for your answer.

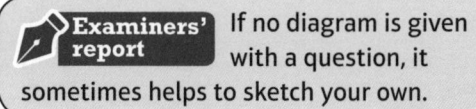

.... **(2 marks)**

Had a go ☐   Nearly there ☐   Nailed it! ☐

GEOMETRY & MEASURES

Line segments

Target grade 4 **Guided**

1. Find the length of the following line segment. Give your answer to 3 significant figures.

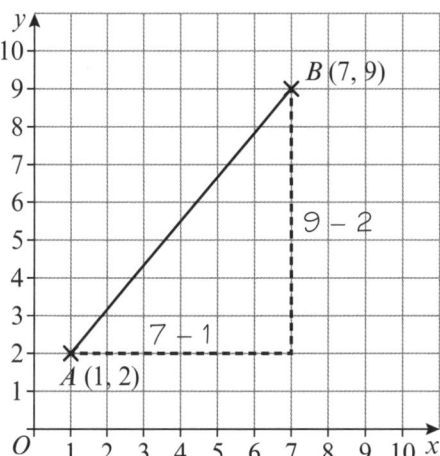

$AB^2 = \ldots\ldots^2 + \ldots\ldots^2$

$AB^2 = \ldots\ldots$

$AB = \sqrt{\ldots\ldots}$

$AB = \ldots\ldots$

Draw a horizontal and a vertical line to make a triangle.

(2 marks)

Target grade 4 **Guided**

2. Find the length of the following line segment. Give your answer to 3 significant figures.

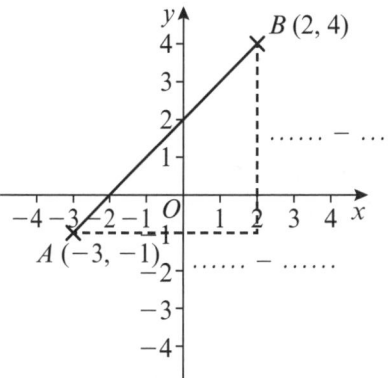

$AB^2 = \ldots\ldots^2 + \ldots\ldots^2$

$AB^2 = \ldots\ldots$

$AB = \sqrt{\ldots\ldots}$

$AB = \ldots\ldots$

There is no grid so use subtraction to work out the distance across and the distance down.

(2 marks)

Target grade 4

3. Point A has coordinates $(3, 1)$ and point B has coordinates $(11, 7)$. Work out the length of the line segment AB.

Problem solved! Draw a sketch showing both points in roughly the right positions.

.... **(2 marks)**

Target grade 4

4. The points $A(-2, -6)$ and $B(4, 2)$ are the opposite ends of a diameter of a circle.

(a) Find the coordinates of the centre of the circle.

.... **(2 marks)**

(b) Show that the radius of the circle is 5 units.

.... **(2 marks)**

91

Trigonometry 1

Had a go ☐ **Nearly there** ☐ **Nailed it!** ☐

1 Work out the size of each of the angles marked with letters.
   Give each answer to 3 significant figures.

   SOH CAH TOA

   (a)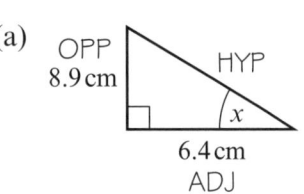

   $\tan x = \dfrac{\text{opp}}{\text{adj}} = \dfrac{\rule{2cm}{0.4pt}}{\rule{2cm}{0.4pt}}$

   $x = \tan^{-1} \rule{3cm}{0.4pt}$

   $x = \rule{2cm}{0.4pt} °$ **(3 marks)**

   Start by labelling the sides of the triangle. Then write down the trigonometric ratio that uses these two sides.

   (b)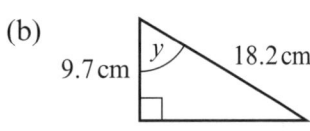

   $\rule{2cm}{0.4pt} y = \dfrac{\text{adj}}{\text{hyp}} = \dfrac{\rule{2cm}{0.4pt}}{\rule{2cm}{0.4pt}}$

   $y = \rule{2cm}{0.4pt}^{-1} \rule{2cm}{0.4pt}$

   $y = \rule{2cm}{0.4pt} °$ **(3 marks)**

2 One end of a rope is tied to the top of a vertical mast of height 7.2 m.
   When the rope is pulled tight, the other end is on the ground 3.7 m from the base of the mast. Work out the angle between the ground and the rope.

   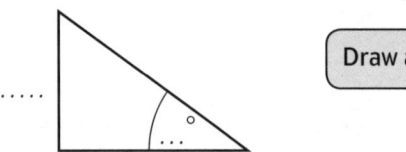

   Draw a diagram.

   ....° **(3 marks)**

3 The diagram shows two right-angled triangles.
   Work out the size of angle x.
   Give your answer correct to 3 significant figures.

   You will have to use Pythagoras' theorem on the bottom triangle first.

   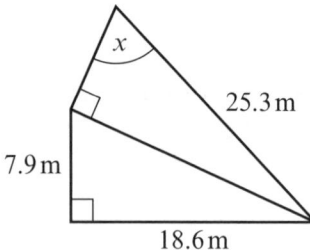

   ....° **(2 marks)**

4 The diagram shows a pitched roof.
   Hayley wants to use smooth tiles to cover the roof.
   The smooth tiles can only be used when the angle, x, is no more than 17°.
   Can she use the smooth tiles on her roof?
   Give a reason for your answer.

   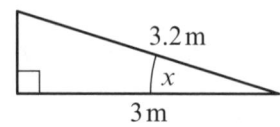

   .... **(3 marks)**

Had a go ☐ **Nearly there** ☐ **Nailed it!** ☐

GEOMETRY & MEASURES

Trigonometry 2

Tricky Topic

Target grade 5

Guided

5 Work out the length, in cm, of each of the marked sides.
   Give each answer correct to 3 significant figures.

   SOH CAH TOA

   (a)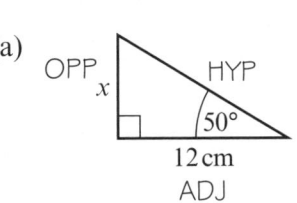

   tan ....° = $\dfrac{\text{opp}}{\text{adj}}$ = $\dfrac{x}{..............}$

   x = ... × tan 50°

   x = ... cm   **(3 marks)**

   Start by labelling the sides of the triangle. Then write down the trigonometric ratio that uses the given and unknown side.

   (b)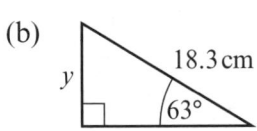

   .... 63° = $\dfrac{\text{opp}}{\text{hyp}}$ = $\dfrac{y}{18.3}$

   y = ... × .... 63°

   y = ... cm   **(3 marks)**

Target grade 5

Guided

6 A ladder is 6 m long. The ladder rests against a vertical wall with the foot of the ladder resting on horizontal ground. The ladder makes an angle of 63° with the ground when it is leaning against the wall.
   How far does the ladder reach up the wall?

   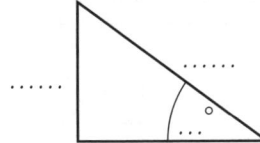

   Draw a diagram.

   .... m   **(2 marks)**

Target grade 5

7 A tower 40 m stands at a point A. At a point B on the ground which is level with the foot of the tower, the angle of elevation of the top of the tower is 36°.
   Work out the distance of B from A.

   .... m   **(2 marks)**

Target grade 5

8 The diagram shows a vertical pole standing on horizontal ground. The points A, B and C are in a straight line on the ground. The point D is at the top of the pole so that DC is vertical. The angle of elevation of D from A is 35°.

   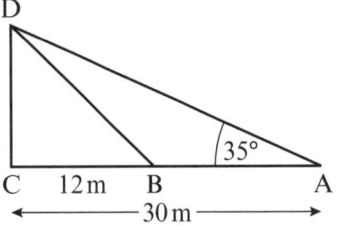

   (a) Work out the height of the pole.
       Give your answer correct to 3 significant figures.

   .... m   **(2 marks)**

   (b) Work out the size of the angle of elevation of D from B.
       Give your answer correct to 3 significant figures.

   .... °   **(2 marks)**

93

GEOMETRY & MEASURES

Had a go ☐ Nearly there ☐ Nailed it! ☐

Exact trigonometry values

1 Complete the table.

> You must remember these for the exam.

	0°	30°	45°	60°	90°
sin	....	$\frac{1}{2}$	....	....	....
cos	....	....	$\frac{1}{\sqrt{2}}$	....	....
tan	....	....	....	$\sqrt{3}$	....

(5 marks)

2 Work out the length, in cm, of the side marked x.

SOH CAH TOA

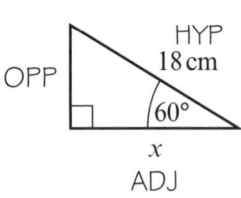

$\cos 60° = \dfrac{\text{adj}}{\text{hyp}} = \dfrac{..................}{..................}$

$x = ... \times \cos 60°$

$x = ...$ cm

(3 marks)

> Start by labelling the sides of the triangle. Then write down the trig ratio that uses the given and unknown sides.

3 Work out the size of each of the angles marked with letters.

SOH CAH TOA

(a)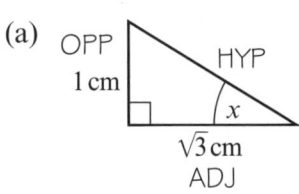

$\tan x = \dfrac{..........}{..........} = \dfrac{..........}{..........}$

$x = \tan^{-1} ....$

$x = ...°$

(3 marks)

(b)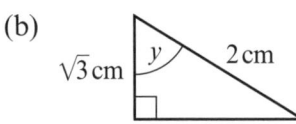

$.................. \, y = \dfrac{..........}{..........} = \dfrac{..........}{..........}$

$y = ...^{-1} ....$

$y = ...°$

(3 marks)

4 Alan is sitting on the ground and his distance from the base of a tower is 30 feet. The angle of elevation from Alan to the top of the tower is 60°.
Work out the height of the tower. Leave your answer as an exact value.

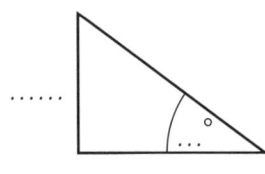

> Draw a diagram.

.... feet (3 marks)

Had a go ☐ Nearly there ☐ Nailed it! ☐   **GEOMETRY & MEASURES**

Measuring and drawing angles

Target grade 1

Guided

1 Measure the sizes of the following angles.   *First estimate the size of the angle.*

(a)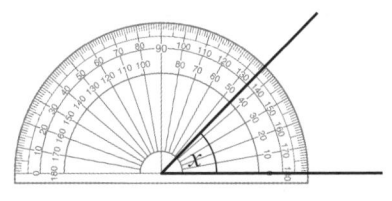

x = ...° **(1 mark)**

(b)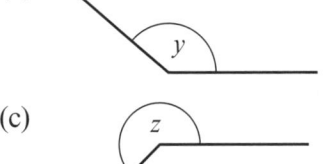

y = ...° **(1 mark)**

(c) 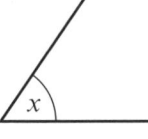   *Measure the small angle and then subtract it from 360°.*

z = ...° **(1 mark)**

Target grade 1

Guided

2 In the space below, accurately draw the following angles.

(a) 74°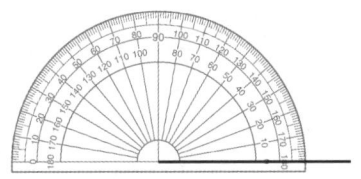

Draw a line and place your protractor at one end. Mark the angle to the nearest degree then draw a straight line between the end of your line and the mark.

(1 mark)

(b) 148°   (c) 262°

(1 mark)   **(1 mark)**

Target grade 1

3 Measure and name the following angles.   *To name the angles, choose from acute, obtuse and reflex.*

(a) 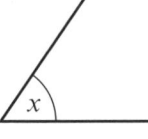   (b) 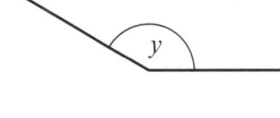   (c)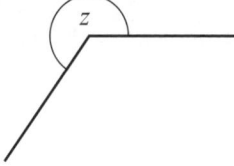

x = ... **(1 mark)**   y = ... **(1 mark)**   z = ... **(1 mark)**

Target grade 1

4 Make an accurate drawing of the following triangle.

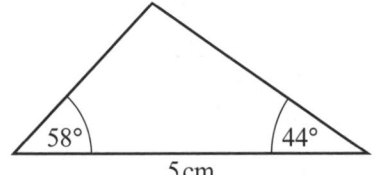

(2 marks)

GEOMETRY & MEASURES

Had a go ☐ Nearly there ☐ Nailed it! ☐

Measuring lines

Target grade 1

1 Measure the lengths of these lines. State the units of measurement.

   First estimate the length of the line.

   (a)

   (b)

   .... (1 mark)   .... (1 mark)

   (c) ─────────────

   .... (1 mark)

Target grade 1

2 In the space below, draw straight lines with the following lengths.

   (a) 52 mm   (b) 6 cm   (c) 7.8 cm

   (1 mark)   (1 mark)   (1 mark)

Target grade 1

3 Mark the mid-point of the line AB with a cross (×).

   Measure the line and then divide by 2.

   Guided

   A ──────────────────── B (2 marks)

Target grade 1

4 The diagram shows an adult woman standing next to a building.
   The woman and the building are drawn to the same scale.
   Work out an estimate for the height, in metres, of the building.

   1.6 m is a good estimate for the height of an adult woman.

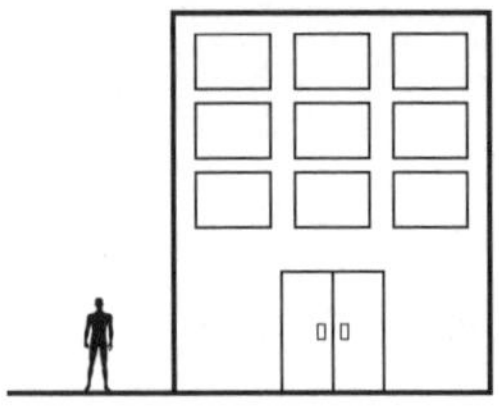

   .... m (3 marks)

Target grade 1

5 The picture shows a house. 
   (a) Write down an estimate for the height, in metres, of one of the front doors.

   .... m   (1 mark)

   (b) Using your answer for part (a), work out an estimate, in metres, for the height of the house.

   .... m (1 mark)

Had a go ☐   Nearly there ☐   Nailed it! ☐

Plans and elevations

1 Match each solid shape to its net.

When the net (arrowed) is folded it will be a cube.

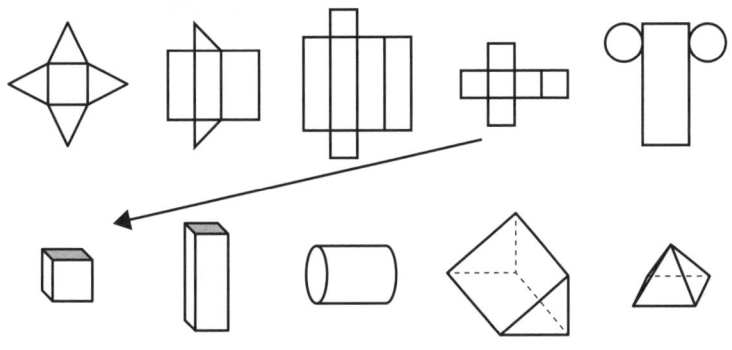

(3 marks)

2 The diagram shows a solid object made of eight identical cubes.

(a) On the grid below, draw the elevation of the solid object from the direction of the arrow

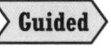

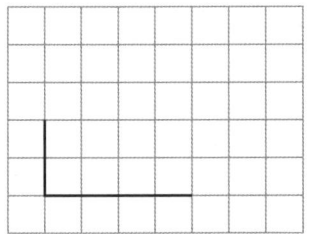 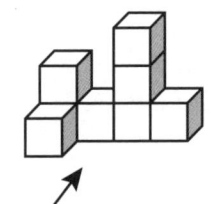

 Use lines to show changes in height or depth.

(2 marks)

(b) On the grid below, draw the plan of the solid object.

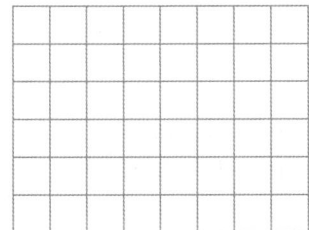

(2 marks)

3 Here are the plan, front elevation and side elevation of a 3-D shape.

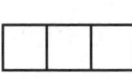 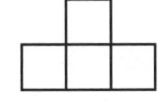 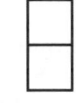

plan   front elevation   side elevation

In the space below, draw a sketch of the 3-D shape.

(2 marks)

4 Here are the plan and front elevation of a solid shape. On the third grid, draw the side elevation of the solid shape.

 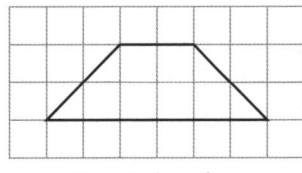 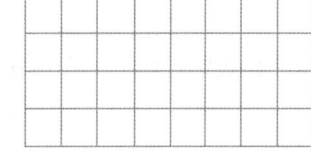

Plan   Front elevation

(2 marks)

97

GEOMETRY & MEASURES

Had a go ☐ Nearly there ☐ Nailed it! ☐

Scale drawings and maps

Target grade 3 **Guided**

1 The lines below are drawn to scale.
Work out the actual lengths by using the scales.

Examiners' report: Measure lines to the nearest mm.

(a) 1 cm to 10 m

Actual length = ... × 10 m = ... m **(2 marks)**

(b) 1 cm to 5 km

(c) 1 cm to 15 km

.... km **(2 marks)**   .... km **(2 marks)**

Target grade 3 **Guided**

2 On a map the distance between two towns is measured and recorded.
Work out the actual distance between the towns using the scale below.

(a) Distance on map = 5 cm
Scale = 1 : 50 000

Actual distance = ... × 50 000 = ... cm

= .... ÷ 100 = ... m

= .... ÷ 1000 = ... km **(2 marks)**

(b) Distance on map = 12 cm
Scale = 1 : 100 000

(c) Distance on map = 15.4 cm
Scale = 1 : 1 000 000

.... km **(2 marks)**   .... km **(2 marks)**

Target grade 3

3 What distance on a map will represent an actual distance of

Convert 10 km into cm.

(a) 10 km using a scale
1 : 50 000

(b) 15 km using a scale
1 : 100 000

(c) 50 km using a scale
1 : 1 000 000?

.... cm **(2 marks)**   .... cm **(2 marks)**   .... cm **(2 marks)**

Target grade 3

4 Arthur uses a scale of 1 : 300 to make a model of an aeroplane.

(a) The wing length of the model is 5 cm. Work out the wing length of the aeroplane.

.... m **(2 marks)**

(b) The length of the aeroplane is 45 m. Work out the length of the model.

.... cm **(2 marks)**

Had a go ☐ Nearly there ☐ Nailed it! ☐

GEOMETRY & MEASURES

Constructions 1

1 Use a ruler and compasses to construct the perpendicular bisector of AB.

 Guided

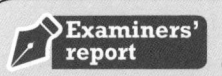 **Examiners' report** Do not rub out the arcs you make when using your compasses.

1. Draw an arc, centre A with radius more than half the length of AB, above and below the line segment AB.
2. Draw another arc, centre B, with the same radius, above and below the line segment AB.
3. Draw a line through the two points where the arcs cross each other above and below the line segment AB.

(2 marks)

2 Use a ruler and compasses to construct the perpendicular to the line segment AB that passes through the point T.

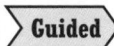 Guided

You must show all your construction lines.

Draw two arcs, centre T, with the same radius to cross either side of T.

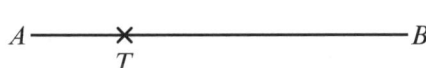

(2 marks)

3 Use a ruler and compasses to construct the perpendicular from P to the line segment AB.

× P

A ——————— B

(3 marks)

99

GEOMETRY & MEASURES Had a go ☐   Nearly there ☐   Nailed it! ☐

Constructions 2

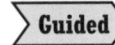 **Target grade 3**

Guided

4 Use a ruler and compasses to construct a triangle with sides of lengths 3.5 cm, 4 cm and 5 cm.

> Do not rub out the arcs you draw when using your compasses.

(2 marks)

> 1. Draw a horizontal line of 5 cm and label it *AB*.
> 2. Set the compasses at 3.5 cm then draw an arc with centre *A*.
> 3. Set the compasses at 4 cm then draw the arc with centre *B*.
> 4. Draw lines from the point of intersection to *A* and *B*.

 **Target grade 4**

Guided

5 Use a ruler and compasses to construct the bisector of angle *ABC*.

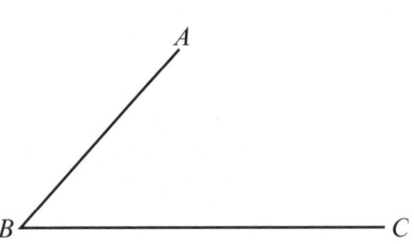

> 1. Draw an arc, centre *B*, to cross *AB* at *P* and *BC* at *Q*.
> 2. Draw an arc, centre *P*, and an arc, centre *Q*, with the same radius. The two arcs intersect.
> 3. Draw a line through point *B* and the point of intersection.

(2 marks)

 **Target grade 4**

6 Use a ruler and compasses to construct a 60° angle at *A*.
You must show all your construction lines.

A ——— *B*

(2 marks)

 **Target grade 4**

7 Use a ruler and compasses to construct a 45° angle at *A*.
You must show all your construction lines.

A ——— *B*

(2 marks)

100

Had a go ☐   Nearly there ☐   Nailed it! ☐

GEOMETRY & MEASURES

Loci

Target grade 4

Guided

1   Draw the locus of all points that are exactly 2 cm from the line AB.

1. Draw a circle of radius 2 cm with centre A.
2. Draw a circle of radius 2 cm with centre B.
3. Draw two parallel lines 2 cm above and below the line AB.

(2 marks)

Target grade 4

Guided

2   The diagram shows the boundary of a rectangular garden, $ABCD$.
A dog is tied to corner B with a rope of length 6 m.
Shade the region where the dog can reach.

1 cm represents 2 m.

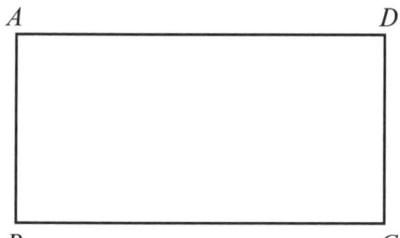

1. Use the scale to set the compasses at the required distance.
2. Draw an arc with centre B.
3. Shade the required region.

(3 marks)

Target grade 4

3   P, Q and R represent three radio masts on a plan. Signals from mast P can be received 100 km away, from mast Q 50 km away and from mast R 75 km away. Show by shading, the region in which signals can be received from all three masts.

1 cm represents 25 km.

× Q

P ×

× R

(3 marks)

Target grade 4

4   ABC is a triangle. Shade the region inside the triangle which is both less than 3 cm from the point B and closer to line AC than AB.

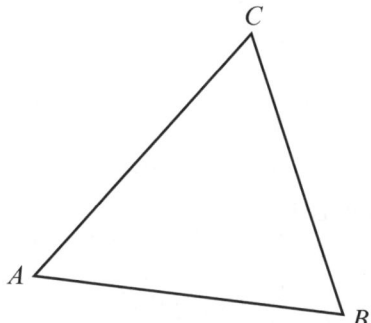

(4 marks)

101

Bearings

1 Work out the bearing of

(a)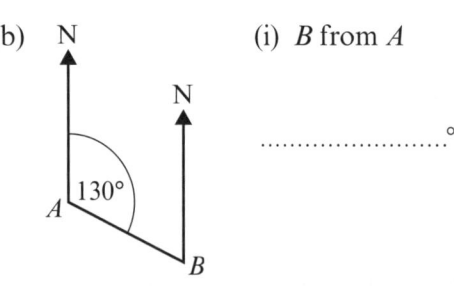

    (i) *B* from *A*

    ....°

    (ii) *A* from *B*

    180° + ...°

    = ...°

(2 marks)

(b)

    (i) *B* from *A*

    ....°

    (ii) *A* from *B*.

    ....°

(2 marks)

2 Draw a line on a bearing of

(a) 30° 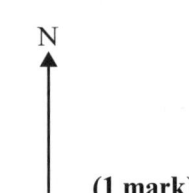 **(1 mark)**

(b) 200°  **(1 mark)**

(c) 320° **(1 mark)**

3 The diagram shows three locations on a map. The scale of the map is 1 cm to 4 km.

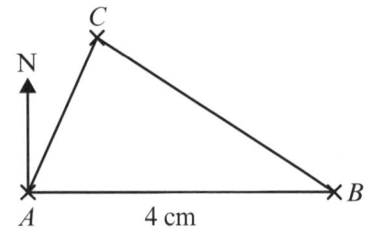

Problem solved! You will need to have a ruler and a protractor with you in your exam. Measure the angle of *AC* clockwise from north to find the bearing of *C* from *A*.

(a) Find the actual distance between *A* and *B*. ....km **(1 mark)**

(b) Measure the bearing of *C* from *A*. ....° **(1 mark)**

D is a fourth location. The actual distance of *D* from *A* is 16 km.
The bearing of *D* from *A* is 120°.

(c) Mark with a cross (×) the position of *D* on the diagram. Label point *D*. **(2 marks)**

Had a go ☐ Nearly there ☐ Nailed it! ☐   **GEOMETRY & MEASURES**

Circles

Target grade 1 **Guided**

1 Label the diagram with the correct names for the parts of a circle.

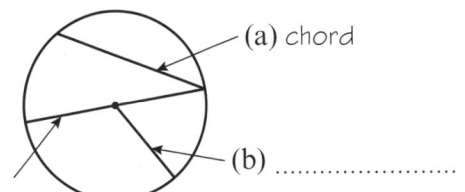
(a) chord
(b) ....
(c) ....
(3 marks)

Target grade 3 **Guided**

2 Work out the circumferences of the following circles. Give your answers correct to 3 significant figures.

You need to learn the formula for the circumference of a circle.

(a) 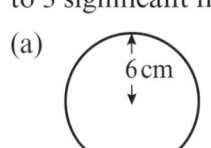
$C = 2 \times \pi \times r$

$= 2 \times \pi \times \text{..........................}$

$= \text{..........................} \text{ cm}$ **(2 marks)**

(b)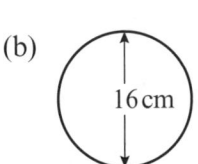

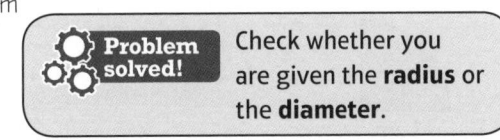
Problem solved! Check whether you are given the **radius** or the **diameter**.

..........................cm **(2 marks)**

Target grade 3 **Guided**

3 Work out the radii of circles with the following circumferences. Give your answers correct to 3 significant figures.

(a) circumference = 35 cm

$C = 2 \times \pi \times r$

.... = $2 \times \pi \times r$

$r = $ .... ÷ ....

radius = ... cm **(2 marks)**

(b) circumference = 92 cm

radius = ... cm **(2 marks)**

Target grade 4

4 Work out the perimeters of the following shapes. Give your answers correct to 3 significant figures.

(a)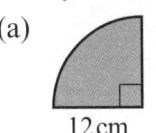
12 cm

.... cm **(3 marks)**

(b)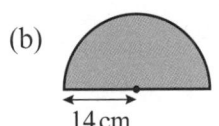
14 cm

.... cm **(3 marks)**

Target grade 4

5 A reel of thread has a radius of 2.5 cm. The thread is wrapped round the reel 200 times. Work out the length of the thread. Give your answer correct to 3 significant figures.

.... cm **(3 marks)**

103

Area of a circle

1 Work out the areas of the following circles.
Give your answers correct to 3 significant figures.

(a) 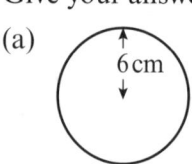 $A = \pi \times r^2$

$= \pi \times \ldots\ldots\ldots^2$

$= \ldots\ldots\ldots$ cm^2

You need to learn this formula.

First work out the radius.

(2 marks)

(b)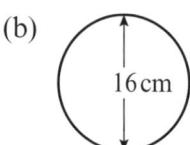

.... cm^2 **(3 marks)**

(c)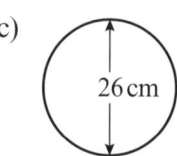

.... cm^2 **(3 marks)**

2 Work out the areas of the following shapes.
Give your answers correct to 3 significant figures.

(a) 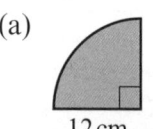 Area of whole circle $= \pi \times r^2 = \pi \times \ldots\ldots \times \ldots\ldots$

$= \ldots\ldots$ cm^2

Area $= \ldots\ldots \div 4 = \ldots\ldots$ cm^2 **(3 marks)**

(b)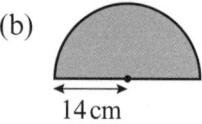

.... cm^2 **(3 marks)**

(c)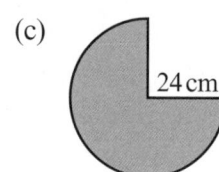

.... cm^2 **(3 marks)**

3 Work out the shaded area of each shape.
Give your answers correct to 3 significant figures.

(a)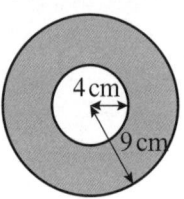

(b)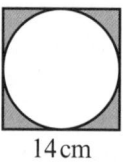

(c)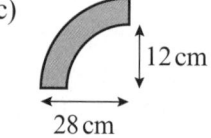

.... cm^2 **(3 marks)** .... cm^2 **(3 marks)** .... cm^2 **(3 marks)**

4 The diagrams show two identical squares.

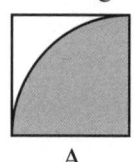 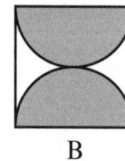

A   B

Diagram A shows a quarter of a circle shaded inside the square.
Diagram B shows two identical semicircles shaded inside the square.

Show that the area of the region shaded in diagram A is equal to area of the region shaded in diagram B.

(3 marks)

Had a go ☐ Nearly there ☐ Nailed it! ☐

GEOMETRY & MEASURES

Sectors of circles

Target grade 5

Guided

1 Work out the arc lengths of the following sectors of circles.
   Give your answers correct to 3 significant figures.

(a)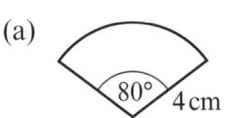

$L = 2 \times \pi \times r \times \dfrac{\ldots\ldots}{360}$

$= 2 \times \pi \times \ldots\ldots \times \dfrac{\ldots\ldots}{360}$

$= \ldots\ldots\ldots$ cm

(3 marks)

You are only calculating the curved length in this question.

(b)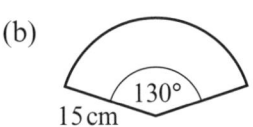

.... cm (3 marks)

(c)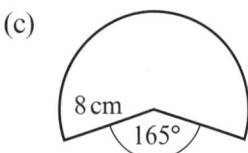

.... cm (3 marks)

Target grade 5

2 Work out the perimeters of the following sectors of circles.
   Give your answers correct to 3 significant figures.

Find the arc length and then add the two radii.

(a)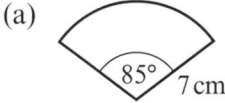

.... cm (4 marks)

(b)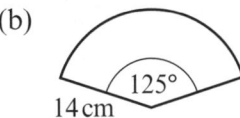

.... cm (4 marks)

(c)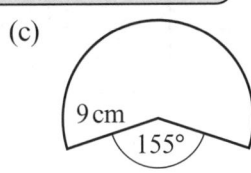

.... cm (4 marks)

Target grade 5

Guided

3 Work out the areas of the following sectors of circles.
   Give your answers correct to 3 significant figures.

(a)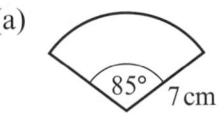

$A = \pi \times r \times r \times \dfrac{\ldots\ldots}{360}$

$= \pi \times \ldots\ldots \times \ldots\ldots \times \dfrac{\ldots\ldots}{360}$

$= \ldots\ldots\ldots$ cm²

(3 marks)

(b)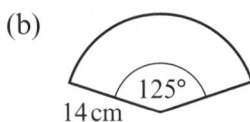

.... cm² (3 marks)

(c)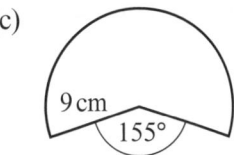

.... cm² (3 marks)

Target grade 5

4 The diagram shows a sector of circle with radius 12 cm.
   Show that the area of the shaded region is 41 cm²
   correct to 2 significant figures.

Examiners' report Label each step of your working to keep track.

(4 marks)

105

GEOMETRY & MEASURES

Had a go ☐ Nearly there ☐ Nailed it! ☐

Cylinders

Target grade 4

Guided

1 Work out the volumes of the following cylinders.
Give your answers correct to 3 significant figures.

(a)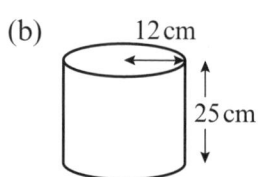

$V = \pi \times r^2 \times h$

$= \pi \times \text{................}^2 \times \text{................}$

$= \text{........................ cm}^3$

(3 marks)

(b)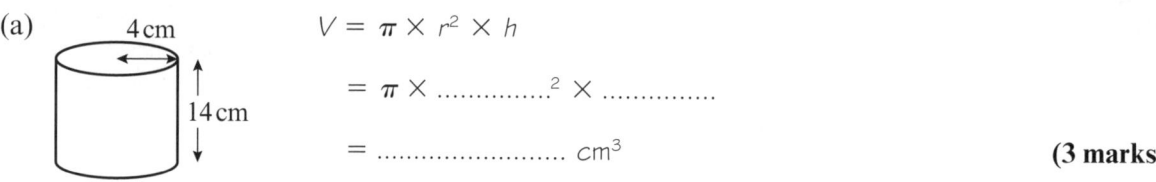

(c)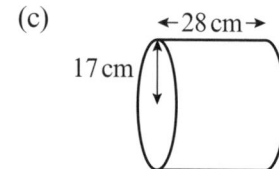

.... cm³ **(3 marks)**   .... cm³ **(3 marks)**

Target grade 4

Guided

2 Work out the total surface areas of the following cylinders.
Give your answers correct to 3 significant figures.

(a)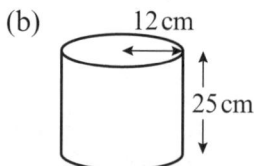

$SA = (2 \times \pi \times r \times h) + (2 \times \pi \times r^2)$

$= (2 \times \pi \times \text{................} \times \text{................}) + (2 \times \pi \times \text{................}^2)$

$= \text{........................ cm}^2$

(3 marks)

(b)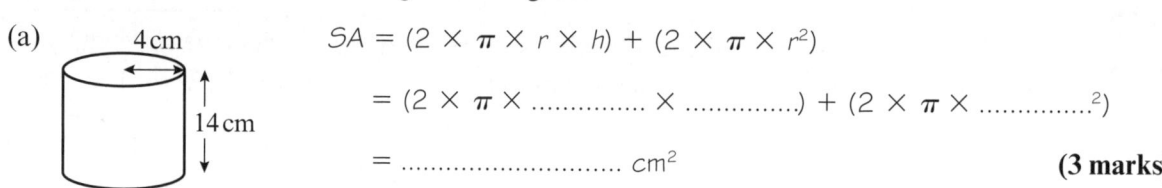

(c)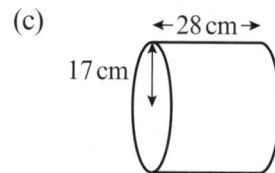

.... cm² **(3 marks)**   .... cm² **(3 marks)**

Target grade 4

3 Diagram A shows a cylinder with radius 15 cm and height 18 cm.
Diagram B shows a cube with side 24 cm.

(a) Show that the volume of the cube is greater than the volume of the cylinder.

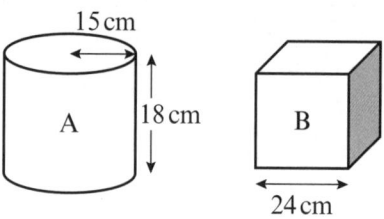

(4 marks)

(b) Which of the shapes has the greater surface area?
You must show your working.

(4 marks)

106

Had a go ☐ Nearly there ☐ Nailed it! ☐

GEOMETRY & MEASURES

Volumes of 3-D shapes

Target grade 5

Guided

1 Work out the volumes of the following shapes.
   Give your answers correct to 3 significant figures.

(a)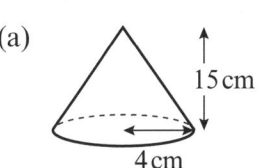

$V = \frac{1}{3} \times \pi \times r^2 \times h$

$= \frac{1}{3} \times \pi \times \ldots\ldots^2 \times \ldots\ldots$

$= \ldots\ldots\ldots$ cm³

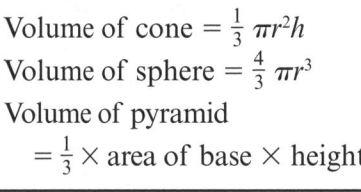

Volume of cone $= \frac{1}{3}\pi r^2 h$
Volume of sphere $= \frac{4}{3}\pi r^3$
Volume of pyramid
$= \frac{1}{3} \times$ area of base $\times$ height

(2 marks)

(b)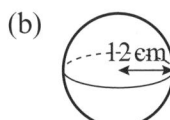

(c)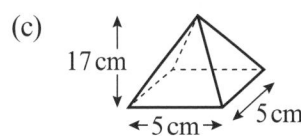

.... cm³ **(2 marks)**   .... cm³ **(2 marks)**

Target grade 5

2 Work out the volumes of the following shapes.
   Give your answers correct to 3 significant figures.

(a)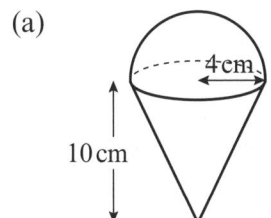

(b)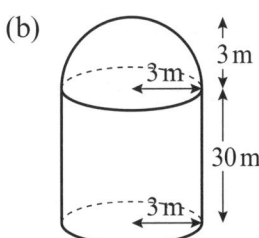

(c)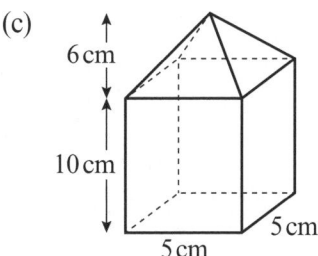

.... cm³ **(3 marks)**   .... m³ **(3 marks)**   .... cm³ **(3 marks)**

Target grade 5

3 A cylinder has radius 4 cm and height 15 cm.
   A cone has base radius 6 cm and height 10 cm.
   Show that the volume of the cylinder is twice
   the volume of the cone.

Examiners' report Calculate both volumes. Then show working for doubling the smaller volume, and write a conclusion.

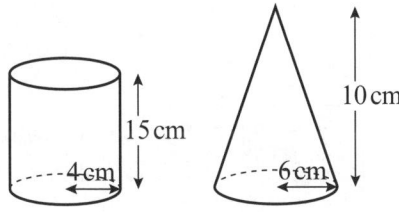

(4 marks)

Target grade 5

4 The diagram shows a cylinder and a sphere.
   The radius of the cylinder is x cm and its height is 9 cm.
   The radius of the sphere is 3 cm.
   The volume of the cylinder is equal to the volume
   of the sphere.
   Show that the radius, x, of the cylinder is 2 cm.

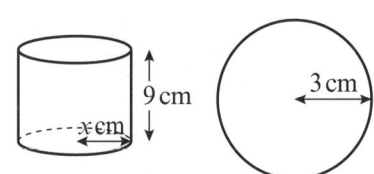

(4 marks)

107

Surface area

1 Work out the surface areas of the following shapes. Give your answers correct to 3 significant figures.

(a)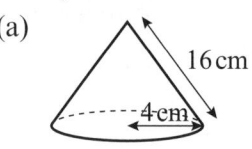

$SA = (\pi \times r^2) + (\pi \times r \times l)$

$= (\pi \times \text{.........}^2) + (\pi \times \text{..........} \times \text{..........})$

$= \text{..............................} \text{ cm}^2$ **(2 marks)**

(b)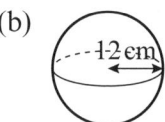

.... cm² **(2 marks)**

(c)

.... cm² **(2 marks)**

2 Work out the surface areas of the following shapes. Give your answers correct to 3 significant figures.

(a)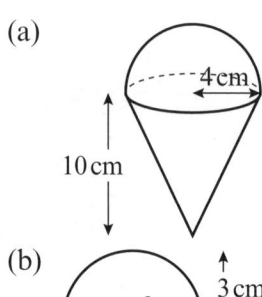

$SA = \text{cone} + \text{hemisphere}$

$= (\pi \times r \times l) + \frac{1}{2}(4 \times \pi \times r^2)$

$= (\pi \times \text{..........} \times \text{..........}) + \frac{1}{2}(4 \times \pi \times \text{..........}^2)$

$= \text{..............................} \text{ cm}^2$ **(3 marks)**

(b)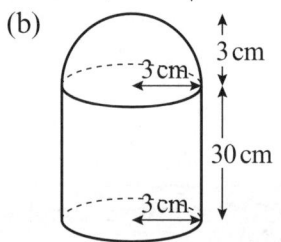

.... cm² **(3 marks)**

(c)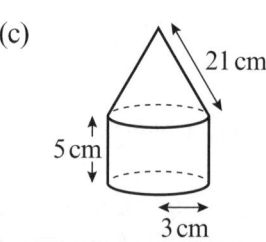

.... cm² **(3 marks)**

3 The diagram shows a cone with vertical height 15 cm and base diameter 16 cm. Work out the curved surface area of the cone.

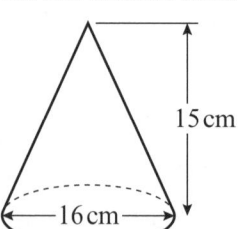

Curved surface area of cone = $\pi r l$

Problem solved! You will need to use Pythagoras' theorem to find the slant height. Sketch the right-angled triangle you need to use.

.... cm² **(4 marks)**

Had a go ☐   Nearly there ☐   Nailed it! ☐

GEOMETRY & MEASURES

Similarity and congruence

Target grade 1
Guided

1   Here are five shapes.

> Congruent shapes have exactly the same size and shape.

(a) Write down the letters of two congruent shapes.

.... and .... **(1 mark)**

> Similar shapes are enlargements of one another.

(b) Write down the letters of two similar shapes.

.... and .... **(1 mark)**

Target grade 1
Guided

2   (a) On the grid below, draw a shape that is congruent to shape A.

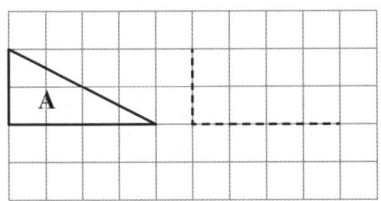

(1 mark)

(b) On the grid below, draw a shape that is similar to shape B.

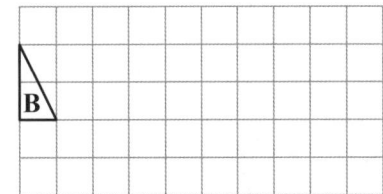

(1 mark)

Target grade 1

3   On each grid, draw a shape that is similar to the shaded shape, but not congruent.

(a)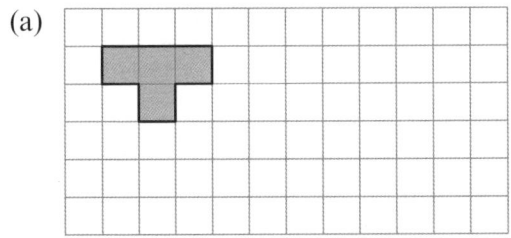

(2 marks)

(b)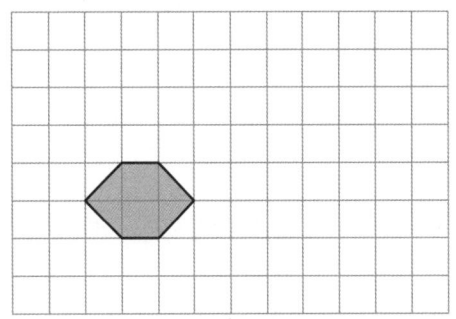

(2 marks)

Target grade 1

4   Here are three shapes.
Ravina says, 'A and C are congruent.'
Anjali says, 'All three are congruent.'
Who is correct?
Explain your answer.

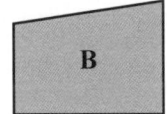

....

(1 mark)

Target grade 2

5   These two cubes are similar.
How many times will the 4 cm cube fit inside the 12 cm cube?

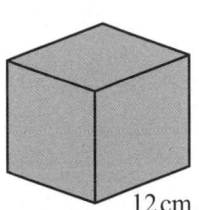

4 cm   12 cm

....

(2 marks)

Similar shapes

1 The two triangles ABC and PQR are mathematically similar.

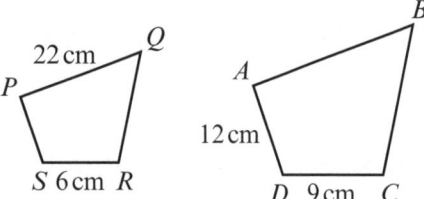

(a) Write down the size of the angle marked x. ....° **(1 mark)**

(b) Work out the length of PR.

$\dfrac{PR}{12} = \dfrac{25}{\text{..............}}$

$PR = \dfrac{25}{\text{..............}} \times \text{..............}$

Use the fact that corresponding sides are in the same ratio.

= .... cm **(2 marks)**

(c) Work out the length of BC.

.... cm **(2 marks)**

2 The diagram shows two quadrilaterals that are mathematically similar.

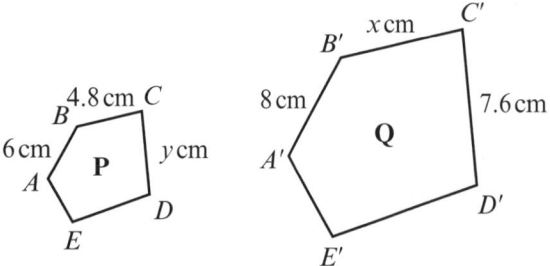

(a) Work out the length of AB.

$\dfrac{AB}{\text{..............}} = \dfrac{\text{..............}}{\text{..............}}$

$AB = \dfrac{\text{..............}}{\text{..............}} \times \text{..............} = \text{..............}$ cm **(2 marks)**

(b) Work out the length of PS.

.... cm **(2 marks)**

3 The diagram shows two pentagons, P and Q, which are mathematically similar.

(a) Work out the value of x.

$x = $ .... cm **(2 marks)**

(b) Work out the value of y.

$y = $ .... cm **(2 marks)**

Had a go ☐   Nearly there ☐   Nailed it! ☐

GEOMETRY & MEASURES

Congruent triangles

Target grade 5

Guided

1 Show that triangle *ABC* is congruent to triangle *DEF*.
You must explain your answer.

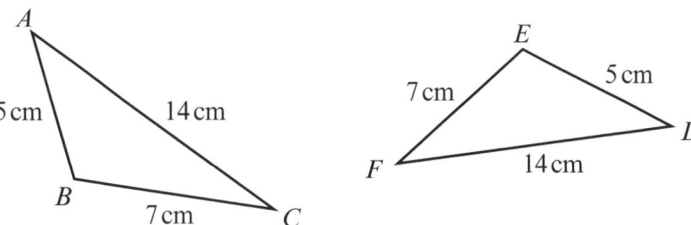

AC = DF

AB = ...

BC = ...

Hence,

.... sides are ....

So the reason is ....   **(3 marks)**

Target grade 5

Guided

2 Show that triangle *ABC* is congruent to triangle *PQR*.
You must explain your answer.

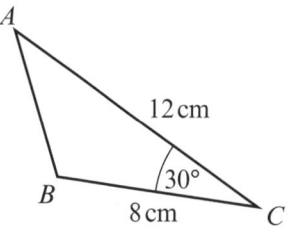 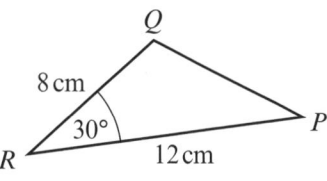

BC = ...

CA = ...

angle *BCA* = angle ....

So the reason is ....   **(3 marks)**

You can write your reasons as SSS, AAS, SAS or RHS.

Target grade 4

3 The diagram shows two triangles.

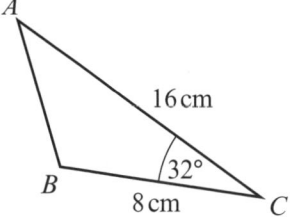 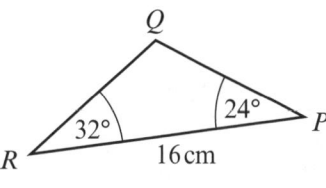

Tick the box with the correct statement. Triangle *ABC* and triangle *PQR*

☐ are definitely congruent   ☐ might be congruent   ☐ are definitely not congruent.

Explain your answer.

....   **(3 marks)**

Vectors

1 Write each vector as a column vector.

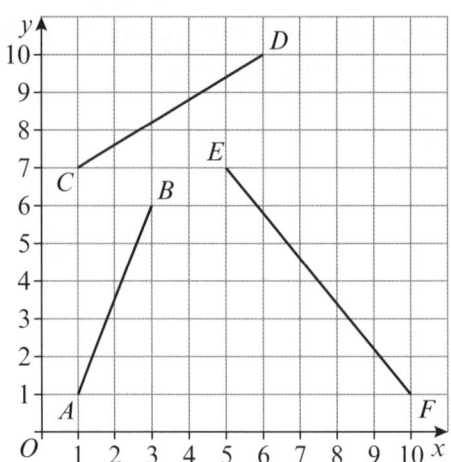

(a) $\vec{AB} = \begin{pmatrix} 2 \\ \ldots \end{pmatrix}$ **(1 mark)**

(b) $\vec{BA} = \begin{pmatrix} \ldots \\ -5 \end{pmatrix}$ **(1 mark)**

(c) $\vec{CD} = \begin{pmatrix} \ldots \\ \ldots \end{pmatrix}$ **(1 mark)**

(d) $\vec{DC} = \begin{pmatrix} \ldots \\ \ldots \end{pmatrix}$ **(1 mark)**

(e) $\vec{EF} = \begin{pmatrix} \ldots \\ \ldots \end{pmatrix}$ **(1 mark)**

(f) $\vec{FE} = \begin{pmatrix} \ldots \\ \ldots \end{pmatrix}$ **(1 mark)**

2 The following diagram shows triangle ABC.
Write down the following vectors in terms of **a** and **b**.

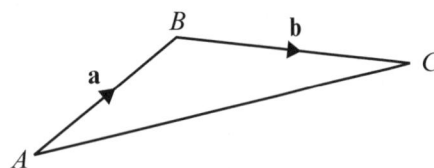

(a) $\vec{AC}$

a + ... **(1 mark)**

(b) $\vec{CA}$

.... **(1 mark)**

3 $ABCD$ is a parallelogram.
AB is parallel to DC.
AD is parallel to BC.
Write down the following vectors in terms of **p** and **q**.

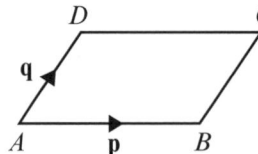

(a) $\vec{AC}$ (b) $\vec{CA}$ (c) $\vec{DB}$ (d) $\vec{BD}$

.... **(1 mark)** .... **(1 mark)** .... **(1 mark)** .... **(1 mark)**

4

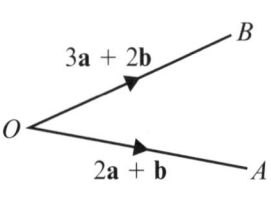

$\vec{OB} = 3\mathbf{a} + 2\mathbf{b}$

$\vec{OA} = 2\mathbf{a} + \mathbf{b}$

Write down the following vectors in terms of **a** and **b**.
Give your answer in its simplest form.

(a) $\vec{AB}$ (b) $\vec{BA}$

.... **(2 marks)** .... **(2 marks)**

Had a go ☐   Nearly there ☐   Nailed it! ☐

GEOMETRY & MEASURES

Problem-solving practice 1

1 *BEG* and *CFG* are straight lines. *ABC* is parallel to *DEF*. Angle *BEF* = 48°.
Angle *BCF* = 30°.

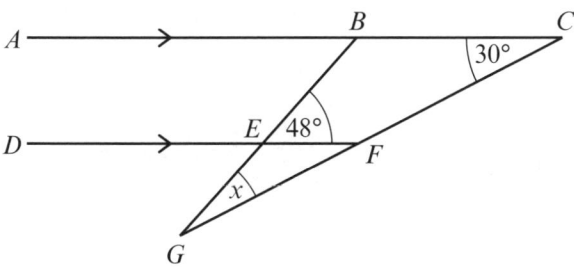

Work out the size of the angle marked *x*.
Give reasons for each step of your working.

....° **(4 marks)**

2 A rectangular tray has length 60 cm, width 40 cm and depth 2 cm.
It is full of water. The water is poured into an empty cylinder of diameter 18 cm and a height of 20 cm. Will there be any water left in the rectangular tray?
You must show all your working.

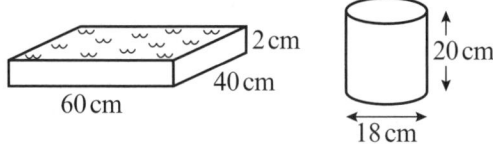

(4 marks)

3

Transformation A is a reflection in the line $y = -x$.
Transformation B is the rotation 180° about (0, 0).

Alison draws a shape and labels it P. She carries out transformation A and labels the image Q. She then transforms Q using transformation B. She labels her final image R.

Describe a single transformation that will map shape P onto shape R.

(3 marks)

113

GEOMETRY & MEASURES

Had a go ☐   Nearly there ☐   Nailed it! ☐

Problem-solving practice 2

Target grade 3

4 The diagram shows a circular garden patio with a radius of 8 m. It has two small circles inside the garden patio. The small circles have a radius of 2 m. Kelly wants to gravel the shaded area. She orders 20 bags of gravel. Each bag covers an area of 9 m². Does she order enough bags to cover the shaded area? You must show all your working.

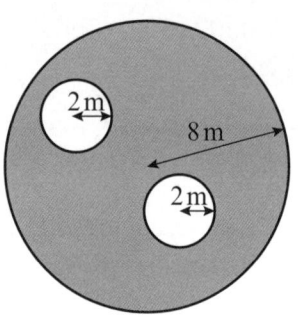

.... (5 marks)

Target grade 4

5 *ABCDEF* is a regular hexagon, *FEHG* and *EXYD* are squares. Angle *HEX* = *x*.
Work out the value of *x*.

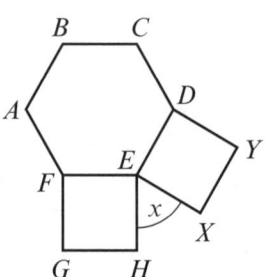

.... (4 marks)

Target grade 5

6 The diagram represents a vertical flagpole, *AB*. The flagpole is supported by two ropes, *BC* and *BD*, fixed to the horizontal ground at *C* and *D*.
AB = 12.8 m, *AC* = 6.8 m and angle *BDA* = 42°.

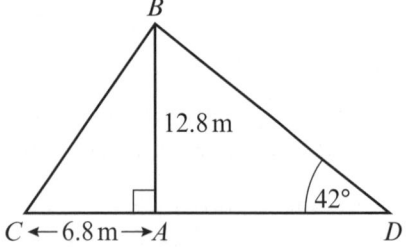

(a) Calculate the size of angle *BCA*.
Give your answer correct to 3 significant figures.

....................................° (2 marks)

(b) Sandeep wants to replace rope *BD*. He buys a new rope of length 20 m.
Is it long enough? You must show your working.

.... (2 marks)

Had a go ☐ Nearly there ☐ Nailed it! ☐

PROBABILITY & STATISTICS

Two-way tables

1 70 children each visited a city last week. The two-way table shows some information about these visits.

> Look for rows or columns with one empty cell.

	Bath	Warwick	Lichfield	Total
Boys	17 − 7 = ...	14	....	32
Girls	7	....	....	70 − 32 = ...
Total	17	25	....	70

Complete the two-way table. **(2 marks)**

2 80 children each chose one school activity from dodgeball, football and rounders. The two-way table shows some information about their choices.

	Dodgeball	Football	Rounders	Total
Girls	12	....	....	41
Boys	18 − 12 = ...	19	....	....
Total	18	....	25	80

(a) Complete the two-way table. **(2 marks)**

(b) How many boys chose football? .... **(1 mark)**

(c) How many girls chose an activity? .... **(1 mark)**

(d) How many girls chose dodgeball? .... **(1 mark)**

3 The two-way table shows some information about the colours of motorbikes and cars in a garage.

	White	Blue	Red	Total
Motorbikes	7	....	....	22
Cars	....	8	....	....
Total	10	17	....	50

(a) Complete the two-way table. **(2 marks)**

(b) Write down the total number of motorbikes. .... **(1 mark)**

(c) How many cars were there in total? .... **(1 mark)**

(d) How many cars were not blue? .... **(1 mark)**

PROBABILITY & STATISTICS

Had a go ☐   Nearly there ☐   Nailed it! ☐

Pictograms

Target grade 1

1. The pictogram shows the numbers of hours of sunshine in Wolverhampton on Monday, Tuesday and Wednesday of one week.

Monday	○ ○ ○ ○
Tuesday	○ ○ ○
Wednesday	○ ○ ◖
Thursday	
Friday	

○ represents 2 hours

(a) Work out the number of hours of sunshine on Monday.

.... hours **(1 mark)**

(b) How many more hours of sunshine were there on Monday than Wednesday?

.... hours **(1 mark)**

There were 4 hours of sunshine on Thursday and 3 hours of sunshine on Friday.

(c) Use this information to complete the pictogram. **(2 marks)**

Target grade 2

2. The pictogram gives information about the number of packets of chocolates sold by a shop some days in one week.

Monday	☐ ☐ ☐ ☐
Tuesday	☐ ☐ ☐
Wednesday	☐ ☐ ▫
Thursday	
Friday	

(a) The total number of packets of chocolates sold on Monday and Tuesday was 130.
Complete the key.

☐ represents .... packets

(1 mark)

(b) How many packets of chocolates were sold on Wednesday?

.... **(1 mark)**

70 packets of chocolates were sold on Thursday.
60 packets of chocolates were sold on Friday.

(c) Use this information to complete the pictogram. **(2 marks)**

Had a go ☐   **Nearly there** ☐   **Nailed it!** ☐

PROBABILITY & STATISTICS

Bar charts

Target grade 1

Guided

1 Shaheen works at an animal shelter for dogs. She has alsatians, bulldogs, labradors and poodles.
   This bar chart shows some information about the alsatians and bulldogs.

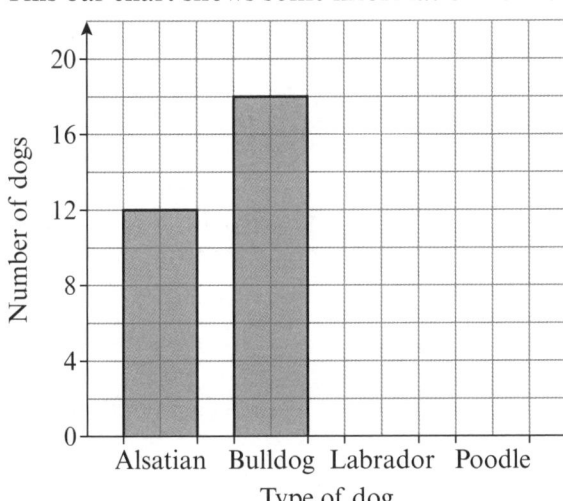

(a) Shaheen also has 8 labradors and 11 poodles in the animal shelter.
Complete the bar chart. **(2 marks)**

(b) Write down the most common dog.

> Look for the highest bar.

.... **(1 mark)**

(c) Work out the total number of dogs in the animal shelter.

12 + ... + ... + ... = ... **(2 marks)**

Target grade 1

Guided

2 Some students each sat an algebra test and a geometry test. Each test was out of 15 marks. The dual bar chart shows the results of four of these students.

(a) Who got more marks in their algebra test than their geometry test?

.... **(1 mark)**

(b) How many more marks did Arn get in her geometry test than in her algebra test?

10 − .... = ... **(1 mark)**

Enzo got 9 marks in his algebra test and 14 marks in his geometry test.

(c) Show this information on the dual bar chart. **(2 marks)**

Target grade 1

3 Julie asked the students in her class which type of pets they had at home.
The bar chart shows some information about the results from her class.

Write down two things that are wrong with the bar chart.

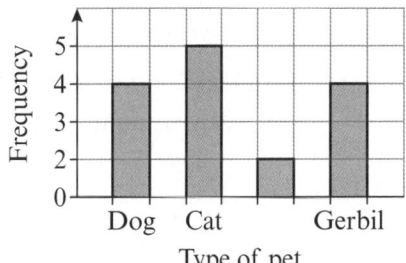

1 ....

2 .... **(2 marks)**

117

PROBABILITY & STATISTICS

Had a go ☐   Nearly there ☐   Nailed it! ☐

Pie charts

1 Brett carries out a survey of 60 people. He asks them their favourite takeaway. The table shows this information. Draw a pie chart to represent this data.

Favourite takeaway	Frequency
Indian	14
Chinese	21
Italian	9
Other	16

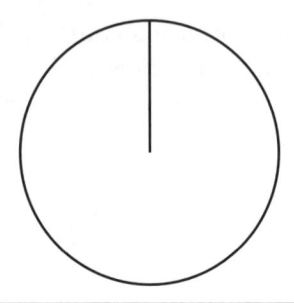

You need to calculate the angles first.

$\frac{14}{60} \times 360° = $ ....

$\frac{\ldots}{60} \times 360° = $ ....

$\frac{\ldots}{60} \times 360° = $ ....

$\frac{\ldots}{60} \times 360° = $ ....

(3 marks)

2 Dhruv asked his friends to tell him their favourite colour. The table shows his results. Draw a pie chart to show his results.

Favourite colour	Frequency
Blue	23
Green	31
Red	22
Yellow	14

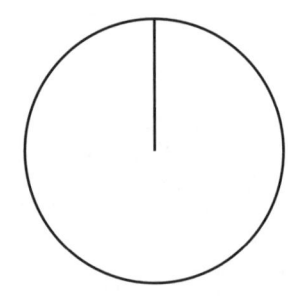

$\frac{\ldots}{90} \times 360° = $ ....

$\frac{\ldots}{90} \times 360° = $ ....

$\frac{\ldots}{90} \times 360° = $ ....

$\frac{\ldots}{90} \times 360° = $ ....

(3 marks)

3 Elaine carries out a survey of some students. The pie chart shows some information about their favourite sport.

(a) 20 students said that cricket is their favourite sport. How many students said that darts is their favourite sport?

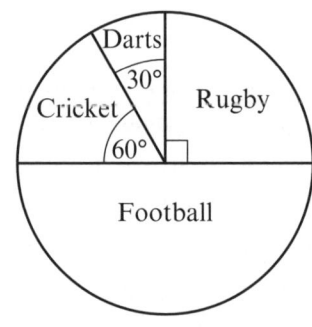

....

(1 mark)

(b) Show that 120 students took part in the survey.

(2 marks)

Had a go ☐   Nearly there ☐   Nailed it! ☐   **PROBABILITY & STATISTICS**

Scatter graphs

1 The weights of seven magazines and the number of pages in each one were recorded. The scatter graph gives information about these results.

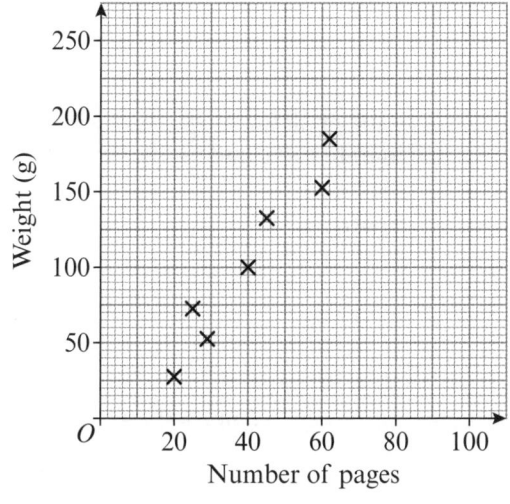

(a) What type of correlation does this scatter graph show?

.... **(2 marks)**

(b) Estimate the weight, in g, of a magazine with 50 pages.

   Draw a line of best fit.

.... g **(2 marks)**

(c) Estimate the weight, in g, of a magazine with 80 pages.

.... g **(1 mark)**

(d) Nik says, 'As the number of pages increases, the books get heavier.'
Does the scatter graph support Nik's statement?

.... **(1 mark)**

(e) Make two comments explaining why your estimate in part (c) might not be accurate.

....

.... **(2 marks)**

2 The scatter graph gives information about the price and age of motorbikes.

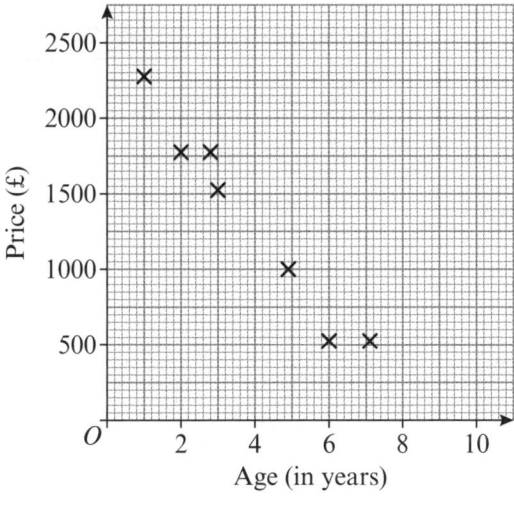

(a) What type of correlation does this scatter graph show?

.... **(2 marks)**

(b) Estimate the price, in £, of a four-year-old motorbike.

£.............................. **(2 marks)**

(c) Comment on the reliability of the estimate in part (b).

....

.... **(1 mark)**

(d) Tim says, 'As the motorbikes get older they get more expensive.'
Does the scatter graph support Tim's statement?

.... **(1 mark)**

PROBABILITY & STATISTICS

Had a go ☐   Nearly there ☐   Nailed it! ☐

Averages and range

1 Here are some numbers:

3, 7, 15, 10, 14, 7, 11

Work out

(a) the mode   (b) the median   (c) the mean   (d) the range.

> The most common number.

> Put the numbers in order and then choose the middle number.

> Sum of all the numbers divided by how many there are.

> Highest − lowest

3, 7, 7, 10, 11, 14, 15

....

(2 marks)   (2 marks)   (2 marks)   (2 marks)

2 The heights, in cm, of five children are shown.

157, 161, 171, 156, 160

(a) Work out the mean height.

(157 + 161 + ... + ... + ...) ÷ 5

= ....cm

(2 marks)

(b) The height of a sixth child is 158 cm.
Work out the new mean.

.... (2 marks)

3 (a) Bob has 3 cards. Each card has a number on it. The numbers are hidden. The mode of the 3 numbers is 6. The mean of the 3 numbers is 7.

> **Problem solved!** If the **mean** of the three numbers is 7, then the **total** of the three numbers must be 3 × 7 = 21.

[?] [?] [?]

Work out the three numbers on the cards.

.... (3 marks)

(b) Emma has 5 cards.
She wants to write down a number on each card such that
  the mode of the 5 numbers is 7
  the median of the 5 numbers is 8
  the mean of the 5 numbers is 9
  the range of the 5 numbers is 5.
Work out the 5 numbers on the cards.

[?] [?] [?] [?] [?]

.... (3 marks)

120

Had a go ☐   Nearly there ☐   Nailed it! ☐

PROBABILITY & STATISTICS

Averages from tables 1

1 The table shows the numbers of goals scored by a football team in each of 30 matches.

Number of goals	Frequency	
0	7	0 × 7 = ...
1	9	1 × 9 = ...
2	6	2 × 6 = ...
3	5	3 × 5 = ...
4	3	4 × 3 = ...

Draw an extra column.

Add up the final column to work out the total number of goals.

Work out

(a) the mode

Mode is .... **(1 mark)**

Examiners' report: The **median** is not necessarily 2 goals. You need to look at the total number of values, and then see where the middle value lies.

(b) the median

Median = $\frac{30 + 1}{2}$ = ...th value = ... **(2 marks)**

(c) the mean

Mean = $\frac{\text{total number of goals}}{\text{frequency}}$ = $\frac{\text{...............}}{\text{...............}}$ = ... **(3 marks)**

(d) the range.

Range = highest value − lowest value = ... − .... = ... **(2 marks)**

2 The table shows information about the results of rolling a dice 25 times. Work out

Score	Frequency
1	3
2	1
3	5
4	3
5	7
6	6

(a) the mode

.... **(1 mark)**

(b) the median

.... **(2 marks)**

(c) the mean.

.... **(3 marks)**

3 Jordan carried out a survey of the number of chocolates 25 students ate in one week.

Number of chocolates	Frequency
0	3
1	4
2	6
3	5
4	4
5	3

(a) Jordan worked out the mean of the number of chocolates eaten. He got an answer of 6. Explain why it is impossible for the mean to be 6.

.... **(1 mark)**

(b) Work out the correct mean.

.... **(2 marks)**

(c) Jordan decides to ask one more person. This person ate no chocolates in this week. Will the mean of the number of chocolates eaten increase or decrease? Give a reason for your answer.

.... **(1 mark)**

PROBABILITY & STATISTICS

Had a go ☐   Nearly there ☐   Nailed it! ☐

Averages from tables 2

4 The table shows information about the number of hours spent on the internet last week.

Number of hours	Frequency f	Mid-point x	$f \times x$
$0 \leq h < 2$	6	1	$6 \times 1 = $ ....
$2 \leq h < 4$	7	3	$7 \times 3 = $ ....
$4 \leq h < 6$	3	5	$3 \times 5 = $ ....
$6 \leq h < 8$	9	7	$9 \times 7 = $ ....
$8 \leq h < 10$	10	9	$10 \times 9 = $ ....

(a) Write down the modal class.

Modal class is .... **(1 mark)**

(b) Write down the class interval which contains the median.

Median $= \dfrac{35 + 1}{2} = $ ....th value.

Median is in class .... **(2 marks)**

(c) Work out an estimate for the mean number of hours.

> Multiply the frequency by the mid-point of each group.

> Add up the final column to work out the total number of hours.

Mean $= \dfrac{\text{total number of hours}}{\text{frequency}} = \dfrac{\text{....................}}{\text{....................}}$

Mean $= $ .... **(4 marks)**

(d) Explain why your answer to part (c) is an estimate.

.... **(1 mark)**

5 Ian asked 25 students how many minutes they each took to get home from school.

Time taken (t minutes)	Frequency
$0 \leq h < 10$	6
$10 \leq h < 20$	7
$20 \leq h < 30$	3
$30 \leq h < 40$	9

(a) Ian used this information to work out the mean of the times taken. He got an answer of 54 minutes. Explain why it is impossible for the mean time to be 54 minutes.

.... **(1 mark)**

(b) Work out an estimate for the mean time taken.

.... **(4 marks)**

(c) Ian realises he has missed out a student. This student takes 32 minutes to get home from school. Ian says, 'The mean time of the students will increase.' Is he correct? Give a reason to support your answer.

.... **(1 mark)**

Had a go ☐   Nearly there ☐   Nailed it! ☐

PROBABILITY & STATISTICS

Line graphs

1 The table shows information about annual turnover of a company in millions of pounds.

Year	Turnover (£ millions)
2008	13
2009	10
2010	12
2011	13
2012	15
2013	16
2014	18

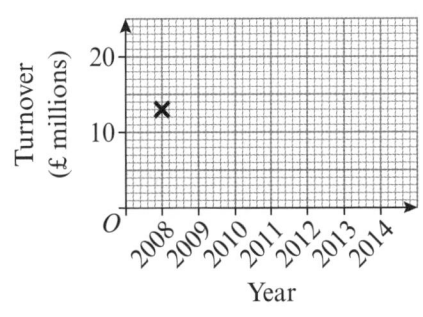

(a) Draw a time series graph to represent this data.

Plot the points from the table.

(2 marks)

(b) Describe the trend.

Use the correct language: upwards or downwards.

....

(1 mark)

2 Joe recorded the number of letters he received each day for a period of time. The graph gives some information about his results.

(a) Write down the modal number of letters.

....

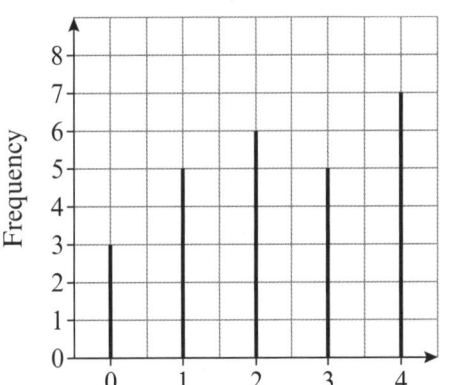

(1 mark)

(b) Work out the total number of letters Joe received during this period of time.

(0 × 3) + (1 × ....) + (2 × ....) + (......... × ....) + (......... × ....)

= ....

(2 marks)

3 The vertical line graph shows the shoe sizes of some children.

(a) Write down the modal shoe size.

Problem solved! Finding the **mean** from a vertical line graph is like finding the mean from a frequency table.

....

(1 mark)

(b) Work out the mean shoe size.

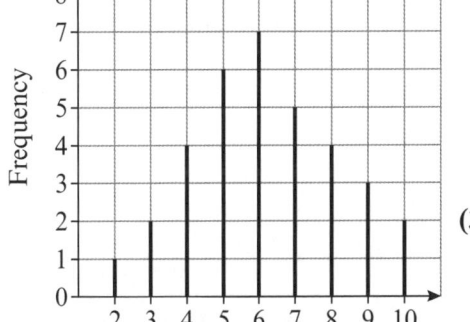

....

(3 marks)

PROBABILITY & STATISTICS

Had a go ☐   Nearly there ☐   Nailed it! ☐

Stem-and-leaf diagrams

Target grade 3

Guided

1 Mary recorded the weights, in kg, of 15 people. Here are her results:

86 50 47 76 73 59 67 79
47 62 51 63 77 61 65

(a) Draw an ordered stem-and-leaf diagram to show this information.

Start by writing down the stems.

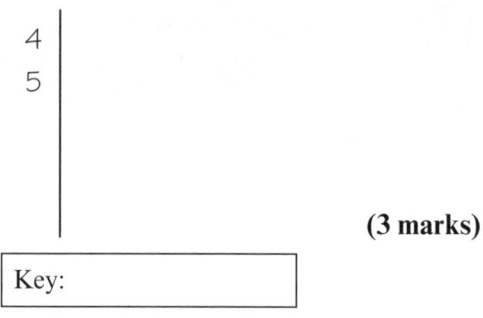

Key:

(3 marks)

(b) Write down the modal weight.

....kg   **(1 mark)**

(c) Work out the median weight.

(d) Show that the range is 39 kg.

.................................kg **(2 marks)**   .... **(2 marks)**

Target grade 3

2 Asha recorded the heart rates of each of 19 people. Here are her results:

53 65 78 82 81 91 59 65 75 97
93 61 72 94 83 79 88 87 65

(a) Draw an ordered stem-and-leaf diagram. **(3 marks)**

(b) Write down the modal heart rate.

Key:

....   **(1 mark)**

(c) Work out the median heart rate.

(d) Work out the range.

.... **(2 marks)**   .... **(2 marks)**

Target grade 3

3 Patrick collected some information about the heights, in cm, of 15 plants. This information is shown in the stem-and-leaf diagram.

Use the key to interpret the stem-and-leaf diagram.

```
6 | 1 2 3 5
7 | 2 4 5 6 7
8 | 1 3 4 8
9 | 3 x
```
Key: 6 | 1 represents 61 cm

(a) Explain why there is no modal value.

.... **(3 marks)**

(b) Work out the median height.

.................................cm **(2 marks)**

(c) The range is 35 cm. Work out the value of x.

.... **(2 marks)**

(d) What plant height does the data value x represent?

.... **(1 mark)**

Had a go ☐ Nearly there ☐ Nailed it! ☐

PROBABILITY & STATISTICS

Sampling

1. Simon wants to find out the number of hours spent on homework in his school each week. He surveys seven children from his class. Here are his results:

   9   7   4   2   6   5   3

   (a) Write down one advantage of taking a sample.

   .... **(1 mark)**

   (b) Use this data to estimate the mean number of hours spent doing homework each week by students.

   > Add up all the values and divide by how many there are.

   .... **(2 marks)**

   (c) Comment on the reliability of this estimate.

   .... **(1 mark)**

   (d) How could Simon reduce bias in his sample?

   .... **(2 marks)**

2. An experiment is carried out by flying paper aeroplanes. The scatter graph shows some information about the distance flown, in m, and the wingspan, in cm. A line of best fit has been drawn.

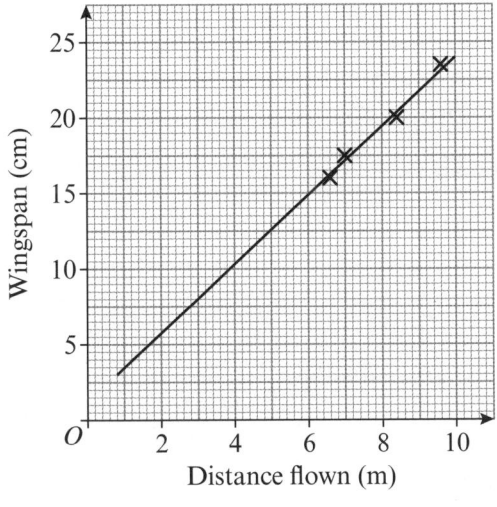

   (a) Use the line of best fit to estimate the wingspan of a plane which flies 9 m.

   > Draw a vertical line to the line of best fit.

   ....cm **(1 mark)**

   (b) Use your line of best fit to estimate the distance flown by a plane with a wingspan of 5 cm.

   ....m **(1 mark)**

   (c) Which of your estimates in part (a) or part (b) is more reliable? Give a reason for your answer.

   .... **(2 marks)**

   (d) Write down one way you could improve this experiment to increase the accuracy of your estimates.

   .... **(1 mark)**

125

PROBABILITY & STATISTICS

Had a go ☐   Nearly there ☐   Nailed it! ☐

Comparing data

1 (a) The following table shows the results of two tests out of 100.

	Mean	Range
Maths	62	14
Statistics	56	20

Compare the test scores in maths and statistics.

Students did better in .... because the mean

was ....

Students' results in maths were more ....

because the range was .... **(2 marks)**

(b) The table shows the amount of rainfall, in mm, in Wolverhampton and Dundee one month.

	Mean	Range
Wolverhampton	25	9
Dundee	39	6

Compare the amount of rainfall in Wolverhampton and Dundee.

....

.... **(2 marks)**

2 Mr Jones kept a record of the number of absences for each student in his class for one term. Here are his results:

   1   0   1   8   6   4   3   5   2   3   4   2

(a) Work out the mean.   (b) Work out the range.

.... **(2 marks)**   .... **(2 marks)**

Mr Singh also kept a record of the number of absences in his class.
The mean number of absences was 5 and the range was 5.

(c) Compare the number of absences for each class.

....

.... **(2 marks)**

3 The exam marks of classes 11A and 11B are shown in the back-to-back stem-and-leaf diagram.

```
            11A       11B
        5 4 1 │ 5 │
      8 6 3 2 │ 6 │ 2 3 4
      8 7 5 4 │ 7 │ 1 3 6 7
              2 │ 8 │ 5 6 8
                │ 9 │ 3
```

Key: 1 | 5 represents 51 marks   Key: 6 | 2 represents 62 marks

Compare the results of class 11A with the results of class 11B.

> Work out the median and the range and then use these values to compare the data.

....

.... **(2 marks)**

Had a go ☐ Nearly there ☐ Nailed it! ☐

PROBABILITY & STATISTICS

Probability 1

Target grade 1

1 On the probability scale, mark with a cross (×) the probability that

(a) it will rain tomorrow

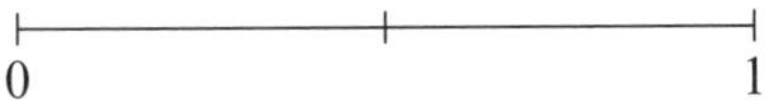

|———|———|
0   1   **(1 mark)**

(b) the sun will not rise tomorrow

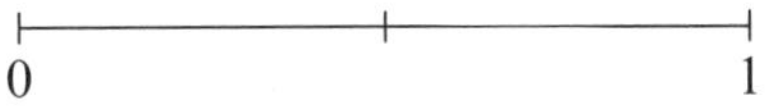

|———|———|
0   1   **(1 mark)**

(c) a coin is tossed and it will land on heads

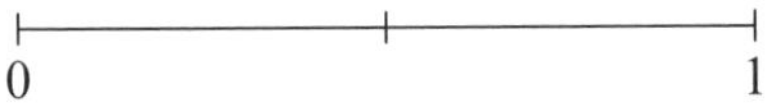

|———|———|
0   1   **(1 mark)**

(d) a dice is rolled and it will land on 6.

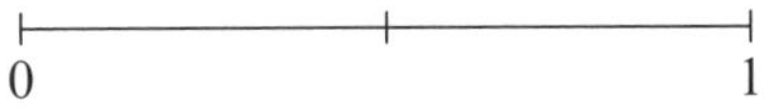

|———|———|
0   1   **(1 mark)**

Target grade 2

Guided

2 John rolls an ordinary dice. The faces are labelled 1, 2, 3, 4, 5 and 6. Write down the probability that he gets

(a) a 5   [How many 5s are there?]

....   **(1 mark)**

(b) an even number   [How many even numbers are there?]

....   **(1 mark)**

(c) a number less than 4

....   **(1 mark)**

(d) a 10.

....   **(1 mark)**

Target grade 1

3 impossible   unlikely   evens   likely   certain

Which word from the above list best describes the likelihood of each of these events?

(a) A dice is rolled and a 7 is shown.

(b) A coin is thrown and lands on tails.

(c) 4 April is the day after 3 April.

.... **(1 mark)**   .... **(1 mark)**   .... **(1 mark)**

Target grade 1

4 (a) The diagram shows a spinner. The spinner can land on A or B or C. Write down the probability that the spinner will land on B.   [How many Bs are there?]

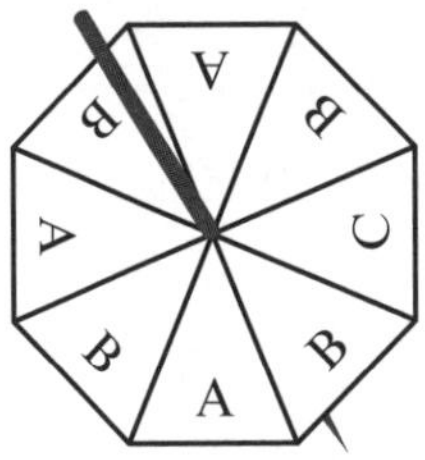

....   **(1 mark)**

(b) Here is a 7-sided spinner. The spinner is spun once. The spinner will land on one of the colours. Write down the probability that the spinner will land on green.

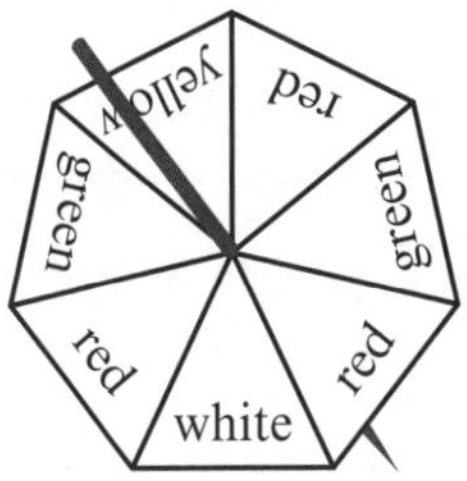

....   **(1 mark)**

127

PROBABILITY & STATISTICS

Had a go ☐   Nearly there ☐   Nailed it! ☐

Probability 2

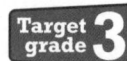

5 A box contains cartons of orange juice, apple juice and mango juice. The table shows each of the probabilities that a carton of juice taken at random from the box will be orange or apple.

Carton of juice	Orange	Apple	Mango
Probability	0.3	0.4	

The probabilities have to add up to 1.

A carton is to be taken at random from the box. Work out the probability that the carton

(a) will be an orange juice or an apple juice

0.3 + ... = ... **(2 marks)**

(b) will be a mango juice.

1 − (.......................... + ...) = ... **(2 marks)**

6 A bag contains counters which are red or green or white or blue. The table shows each of the probabilities that a counter taken at random will be red or green or white.

Colour	Red	Green	White	Blue
Probability	0.35	0.28	0.16	

A counter is to be taken at random from the bag. Work out the probability that the counter will be blue.

1 − (.................... + ... + ...) = ... **(2 marks)**

7 A spinner can land on A, B, C or D. The table shows the probabilities that the spinner will land on each letter B or C or D.

Letter	A	B	C	D
Probability		0.26	0.36	0.17

The spinner is spun once. Work out the probability that the letter on the spinner

(a) will be **B and** C

(b) will be A.

.... **(2 marks)**   .... **(2 marks)**

8 Four athletes Andy, Ben, Carl and Daljit take part in a race. The table shows the probabilities of Andy or Ben winning the race.

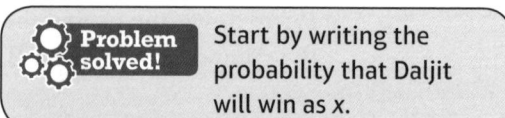

Problem solved! Start by writing the probability that Daljit will win as x.

Athlete	Andy	Ben	Carl	Daljit
Probability	0.3	0.38		

The probability that Carl will win is 3 times the probability that Daljit will win. Work out the probability that the race will be won by

(a) Andy or Ben

(b) Daljit.

.... **(2 marks)**   .... **(2 marks)**

128

Had a go ☐ Nearly there ☐ Nailed it! ☐

PROBABILITY & STATISTICS

Relative frequency

1 The table shows information about the number of orders received each month for six months by an internet company.

Month	Jan	Feb	Mar	Apr	May	Jun
Number of orders	28	63	49	61	53	48

An order is chosen at random.
Work out the probability that the order was received in

(a) May

> First work out the total number of orders.

$\dfrac{53}{\text{\dots\dots}}$

.... **(2 marks)**

(b) January or February or March.

> Add up the numbers for January, February and March.

28 + ... + ... = ...

....
.... **(2 marks)**

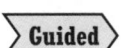

2 The table shows the total scores when Ethan throws three darts 50 times.

Score	1–30	31–60	61–90	91–120	121–150	151–180
Frequency	14	10	9	8	6	3

He throws another three darts. Estimate the probability that he scores

(a) between 31 and 60

.... out of 50 = $\dfrac{\text{\dots\dots}}{50}$ = $\dfrac{\text{\dots\dots}}{5}$ **(1 mark)**

(b) more than 90

.... **(2 marks)**

(c) 120 or fewer.

.... **(2 marks)**

3 A garage keeps records of the cost of repairs it makes to vans. The table gives information about the costs of all repairs which were £500 or less in one month.

Cost (£C)	Frequency
0 < C ≤ 100	20
100 < C ≤ 200	39
200 < C ≤ 300	72
300 < C ≤ 400	33
400 < C ≤ 500	38

(a) Amy needs to repair her van. Estimate the probability her repair costs more than £200.

> **Examiners' report** Don't round any values. The answer is an estimate because it is based on experimental data.

.... **(2 marks)**

(b) Comment on the accuracy of your estimate.

.... **(1 mark)**

PROBABILITY & STATISTICS

Had a go ☐ Nearly there ☐ Nailed it! ☐

Frequency and outcomes

Target grade 1

1. Brett goes to a restaurant.
   He can choose from three types of curry and from three types of naan.
   Brett is going to choose one curry and one naan.
   Write down the probability that he chooses a lamb curry with a butter naan.

Curry	Naan
Chicken	Plain
Lamb	Garlic
Vegetable	Butter

   Label chicken as C, etc.

   (C, ...) (C, ...) (C, ...)

   (L, ...) (L, ...) (L, ...)

   (V, ...) (V, ...) (V, ...)

   Probability = ... **(2 marks)**

Target grade 1

2. Martin is holding three cards, labelled X, Y and Z.
   He mixes them up and then asks Neil to choose a card at random.

   (a) Write down the probability that Neil chooses card Y. .... **(1 mark)**

   Neil replaces his card and Martin mixes the cards up again.
   Martin then asks Tej to choose a card.

   (b) Complete the table of possible outcomes.

Neil's card									
Tej's card									

   (2 marks)

   (c) Work out the probability that Neil and Tej both choose the same card.

   .... **(1 mark)**

   (d) Work out the probability that Neil and Tej choose different cards.

   .... **(1 mark)**

Target grade 3

3. 120 adults were asked if they voted in the general election. 58 of these adults were male.
   7 of the females did not vote. 103 of the adults voted.

   (a) Draw a frequency tree to show this information.

   (3 marks)

   One of the males is chosen at random.

   (b) Work out the probability that this male did not vote.

   .... **(2 marks)**

Target grade 3

4. A bag contains 50 counters. They are all either green or blue.
   A counter is chosen at random. The probability that it is green is $\frac{3}{10}$.
   Work out the number of blue counters in the bag.

   .... **(2 marks)**

Had a go ☐ Nearly there ☐ Nailed it! ☐

PROBABILITY & STATISTICS

Venn diagrams

Target grade 5

Guided

1. The following diagrams represent the subjects studied at college by a group of 30 students. For each diagram

   (a) work out the value of x

   (b) write down the set that x represents.

   (i)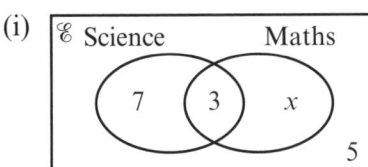

   $30 - (\ldots\ldots\ldots + \ldots\ldots\ldots + \ldots\ldots\ldots)$

   (a) $x = \ldots\ldots\ldots\ldots$ **(1 mark)**

   (b) Students who only study ……………………………… **(1 mark)**

   (ii) 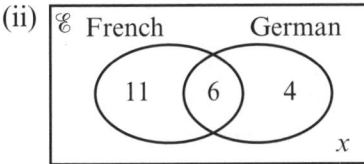   (iii)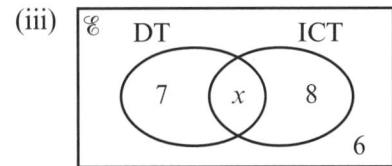

   (a) $x = \ldots\ldots\ldots\ldots$ **(1 mark)**   (a) $x = \ldots\ldots\ldots\ldots$ **(1 mark)**

   (b) ……………………………… **(1 mark)**   (b) ……………………………… **(1 mark)**

2. The Venn diagram shows information about musical instruments played by 40 students. A student is chosen at random. Work out the probability that this student

   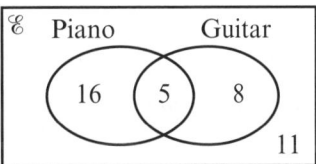

   (a) plays the piano and the guitar

   (b) plays neither instrument

   (c) plays the piano.

   ……………… **(1 mark)**   ……………… **(1 mark)**   ……………… **(1 mark)**

Target grade 5

3. In a class of 30 students, 10 own a PS4, 12 own an Xbox and 4 own both.

   (a) Draw a Venn diagram to represent this information.

   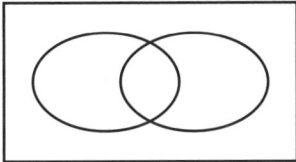

   (2 marks)

   A student is chosen at random. Find the probability that this student

   (b) does not own a PS4 and does not own an Xbox

   ……………………………………… **(2 marks)**

   (c) owns a PS4 or an Xbox but not both.   ……………………………………… **(2 marks)**

131

Set notation

1 $P = \{m, e, t, r, i, c\}$
$Q = \{g, c, s, e\}$
List the members of the set

(a) $P \cap Q$

(b) $P \cup Q$

*$P \cap Q$ means elements that are in **both** P and Q. $P \cup Q$ means elements that are in **either** P or Q.*

.... **(1 mark)**   .... **(1 mark)**

2 $\mathcal{E} = \{\text{positive integers up to 100}\}$
$A = \{\text{multiples of 20}\}$
Dawn says that 200 is a member of A because it is a multiple of 20. Dawn is incorrect. Explain why.

....

.... **(1 mark)**

3 $\mathcal{E} = \{\text{positive whole numbers less than 15}\}$
$X = \{\text{multiples of 3}\}$
$Y = \{5, 6, 7, 8, 9, 10\}$

Examiners' report: There are 14 positive whole numbers less than 15, so make sure that you have written exactly 14 different numbers on your Venn diagram.

(a) Complete the Venn diagram for this information.

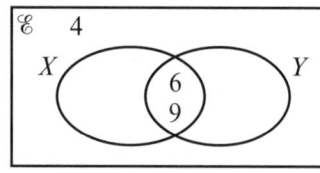

(4 marks)

A number is chosen at random from the universal set $\mathcal{E}$.

(b) What is the probability that it is in the set $X \cup Y$?

.... **(2 marks)**

4 $\mathcal{E} = \{1, 2, 3, 4, 5, 6, 7, 8, 9, 10\}$
$A = \{\text{prime numbers}\}$
$A \cap B = \{3, 5, 7\}$
$A \cup B = \{2, 3, 4, 5, 6, 7\}$

Draw a Venn diagram for this information.

Make sure you label your sets and label the rectangle with the symbol for the universal set $\mathcal{E}$.

(4 marks)

5 Shade in the regions represented by

(a) $P \cup Q$ (b) Q (c) $P \cap Q$

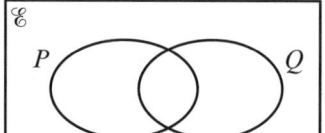 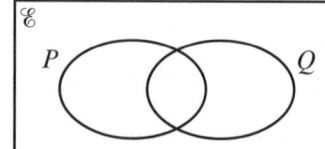 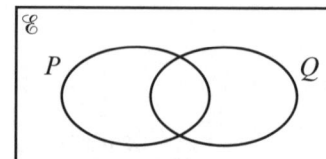

Had a go ☐   Nearly there ☐   Nailed it! ☐

PROBABILITY & STATISTICS

Tricky Topic

Independent events

1 Marcus has 10 counters in a bag.
3 of the counters are yellow and the remaining counters are blue.
Marcus chooses a counter at random and notes the colour.
He then puts the counter back into the bag.
He chooses another counter at random and notes the colour. Work out the probability that

> **Problem solved!** Multiply the probabilities to work out the probability that **both** events happen. For part (c), you can subtract the probabilities that the counters are both yellow or both blue from 1 to find the probability that they are different colours.

(a) both counters will be yellow

$\frac{3}{10} \times \frac{\ldots}{10} = \frac{\ldots}{\ldots}$

(2 marks)

(b) both counters will be blue

$\frac{\ldots}{10} \times \frac{\ldots}{10} = \frac{\ldots}{\ldots}$

(2 marks)

(c) the counters will be different colours.

$1 - \frac{\ldots}{\ldots} - \frac{\ldots}{\ldots} = \frac{\ldots}{\ldots}$

(3 marks)

2 A bag contains 3 blue marbles and 7 red marbles.
A marble is chosen at random, replaced, and then another is taken out.

(a) Complete the probability tree diagram.

First marble / Second marble

- 0.7 → R
  - 0.7 → R
  - ...... → B
- ...... → B
  - 0.7 → R
  - ...... → B

(2 marks)

(b) Work out the probability that one of each colour is chosen.

.... **(3 marks)**

3 Nav and Asha each take a motorcycle test.
The probability that Nav will pass is 0.9.
The probability that Asha will pass is 0.8.

(a) Complete the probability tree diagram.

Nav   Asha

...... Pass

...... Fail

(2 marks)

(b) Work out the probability that only one of them will pass the test.

.... **(3 marks)**

PROBABILITY & STATISTICS

Had a go ☐   Nearly there ☐   Nailed it! ☐

Problem-solving practice 1

Target grade 3

1 80 students each study one of three languages.
The two-way table shows some information about these students.

	French	German	Spanish	Total
Female	15	....	....	39
Male	....	17	....	41
Total	31	28	....	80

(a) Complete the two-way table. **(2 marks)**

(b) One of these students is to be picked at random.
Write down the probability that this student studies French.

.... **(2 marks)**

Target grade 3

2 A school snack bar offers a choice of four snacks. The four snacks are burgers, wraps, fruit and salad. Students can choose one of these four snacks.
The table shows the probability that a student will choose a burger or a wrap.

Snack	Burger	Wrap	Fruit	Salad
Probability	0.25	0.15		

The probability that a student chooses fruit is twice the probability that a student chooses salad.
One student is chosen at random from the students who use the snack bar.

(a) Work out the probability that the student

   (i) did not choose a burger   (ii) chose salad.

.... **(2 marks)**   .... **(2 marks)**

(b) 200 students used the snack bar on Tuesday.
Estimate the number of students who chose a wrap.

.... **(2 marks)**

Target grade 4

3 The heights, in cm, of some plants were measured in Park A and in Park B.
The information is shown in the back-to-back stem-and-leaf diagram.

```
                    Park A       Park B
                           1 | 0 3 4
                   3 3 1 | 2 | 1 2 5 7
Key: 1 | 2 represents 21 cm   6 5 2 0 | 3 | 3 4 6 7   Key: 1 | 0 represents 10 cm
                     7 5 4 | 4 | 2
                         2 | 5 |
```

Compare the heights of plants in Park A with the heights of plants in Park B.

....

.... **(2 marks)**

134

Had a go ☐   Nearly there ☐   Nailed it! ☐

PROBABILITY & STATISTICS

Problem-solving practice 2

4 Heather carries out a survey of 180 Year 11 students.
She asks them their favourite snack.
She draws this accurate pie chart.
Use the pie chart to complete the table.

Favourite snack in Year 11

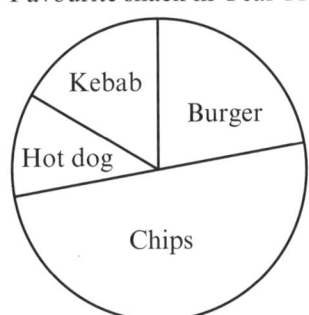

Diagram accurately drawn

Favourite snack in Year 11	Frequency	Angle
Burger	40	....
Chips	90	180°
Hot dog	....	....
Kebab	....	....
Total	180	

(4 marks)

5 Jamie and Rajiv each take an entrance exam.
The probability that Jamie will pass the entrance exam is 0.7.
The probability that Rajiv will pass the entrance exam is 0.75.

(a) Complete the probability tree diagram.

(b) Work out the probability that only one of them will pass the entrance exam.

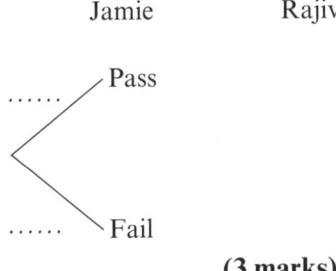

(3 marks)

.... **(3 marks)**

6 In a class of 25 students, 8 study Latin, 10 study Mandarin and 3 study both.

(a) Draw a Venn diagram to represent this information.

(2 marks)

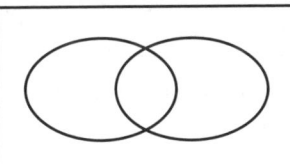

A student is chosen at random.
Find the probability that the student

(b) does not study Latin and does study Mandarin

.... **(2 marks)**

(c) studies Latin or Mandarin but not both.

.... **(2 marks)**

Paper 1
Practice exam paper

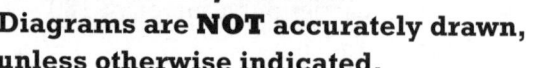

Foundation Tier
Time: 1 hour 30 minutes
Calculators may be used
Diagrams are **NOT** accurately drawn, unless otherwise indicated.
You must **show all your working out**.

1 Sandeep wrote down the temperature at different times on 1 January 2015.

Time of day	Temperature (°C)
3 am	−11
7 am	−5
Noon	7
4 pm	5
8 pm	−3
Midnight	−9

(a) Write down

  (i) the highest temperature

  (ii) the lowest temperature. .... **(2 marks)**

(b) Work out the difference in the temperature between

  (i) 3 am and 7 am

  (ii) noon and 8 pm. .... **(2 marks)**

2 Here is Kate's bank statement for the month of May up to 30 May.

Date	Deposit (£)	Withdrawal (£)	Balance (£)
01/05/15			4240.00
06/05/15		300.00	3940.00
15/05/15	345.00		4285.00
19/05/15		450.00	....
27/05/15	1350.00		....

Kate needs £5400 for a new garden patio.
She needs to pay this money at the end of May.
Does Kate have enough money in her account to pay for the garden patio?

.... **(3 marks)**

3 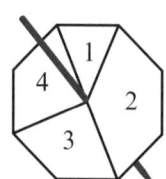 The diagram shows a spinner as an eight-sided shape.
The spinner sections are labelled 1 or 2 or 3 or 4.
The spinner is spun once and lands on a number.

(a) Which number is the most likely?
You must explain your answer. ....

.... **(1 mark)**

(b) Write down the probability of the spinner landing on 1 or 2. .... **(1 mark)**

Ravina spins the spinner once.
Her spinner lands on 2.
Anjali then spins the spinner once.

(c) Has Anjali got a smaller chance, the same chance or a greater chance of getting 2 than Ravina had?
You must explain your answer. ....

.... **(1 mark)**

4 The diagram represents a straight road that joins four villages.

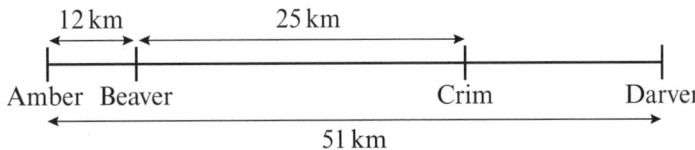

(a) Work out the distance from Crim to Darver.

.... **(1 mark)**

Lewis walks from Amber to Crim.
Harry walks from Beaver to Darver.

(b) Who walks the furthest?
You must explain your answer. ....

.... **(2 marks)**

5 In a survey, 100 male voters and 100 female voters were asked which political party they were voting for in the next general election.
The table below shows information about their replies.

	Conservatives	Labour	Liberal Democrats	Other
Female	28	32	24	16
Male	24	30	18	28

Represent this information in a suitable chart or diagram.

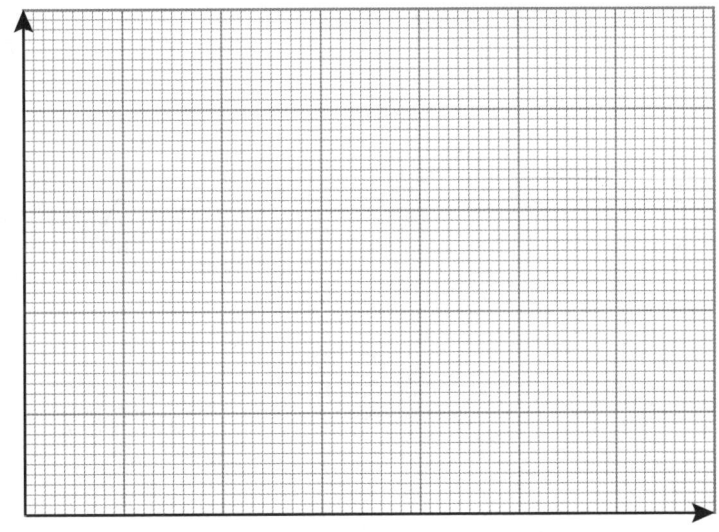

(4 marks)

6 Simplify

(a) $a + a + a + a$ .... **(1 mark)**

(b) $5x + 8x - 3x$ .... **(1 mark)**

(c) $5e - 6f + 7e + 7e - 2f + 4$ .... **(2 marks)**

7 Callum says, $25 - 10 \times 2$ is 30
Len says, $25 - 10 \times 2$ is 5

(a) Who is correct?
Give a reason for your answer. ....

.... **(1 mark)**

(b) Work out the value of $(28 - 4) \div 4 + 10$

.... **(1 mark)**

(c) Put in brackets to make the following calculation correct.

$16 - 4^2 + 3 = -3$ **(1 mark)**

8 Here are two fractions, $\frac{3}{4}$ and $\frac{4}{5}$.
Which is the smaller fraction?
You must show all your working.

.... **(3 marks)**

9 (a) Triangle *ABC* is an equilateral triangle.
Explain why.

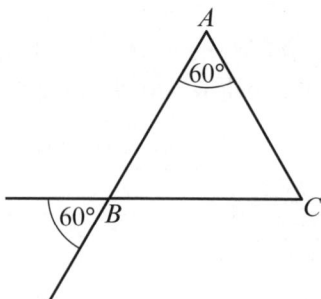

.... **(2 marks)**

PQR is a straight line.
SR = QR.

(b) Work out the value of *x*.
Give reasons for your answer.

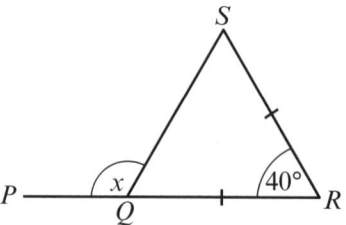

.... **(3 marks)**

10 The diagram shows part of a map. It shows the positions of a statue and a cathedral.

The scale map is 1:10 000
Work out the real distance between the statue and the cathedral.
Give your answers in metres.

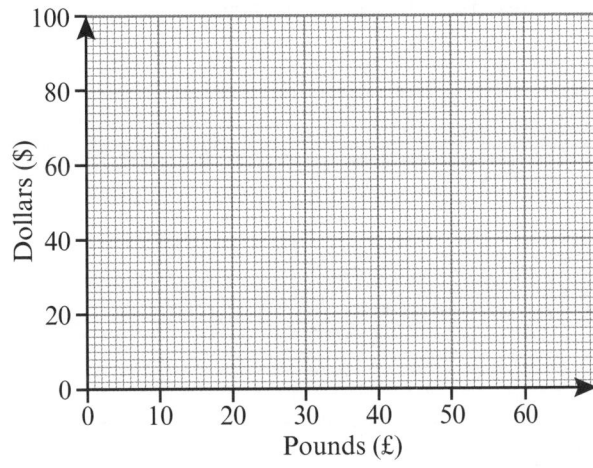

.... **(2 marks)**

11 Factorise

(a) $5x + 10$ .... **(1 mark)**

(b) $x^2 - 6x$ .... **(1 mark)**

12 Pavan comes back from the USA with some money.
Pavan needs to change some money from dollars ($) to pounds (£).

£1 = $1.50

(a) On the grid, draw a conversion graph Pavan can use to change between pounds and dollars.

(2 marks)

Pavan changes 1000 dollars into pounds.

(b) Use your graph to change 1000 dollars into pounds.

.... **(2 marks)**

13 There are $2\frac{1}{2}$ litres of water in a jug.
Laura is going to pour the water into some glasses.
She will fill each glass with 225 ml of water.
Work out the greatest number of glasses she can fill.

.... **(4 marks)**

139

14 Gavin bought his motorbike for £15 000.
The motorbike depreciated by 10% in the first year.
The motorbike depreciated by 15% in the second year.
Show that the value of the motorbike falls below £11 500 after 2 years.
You must show all your working.

(3 marks)

15 This is some information about a class.
There are 40 students in a class.
16 of the students study Latin.
19 of the students study Spanish.
7 of the students study both Latin and Spanish.

(a) Draw a Venn diagram to represent this information.

(4 marks)

(b) Show that 30% of the students do **not** study Latin or Spanish.

(2 marks)

16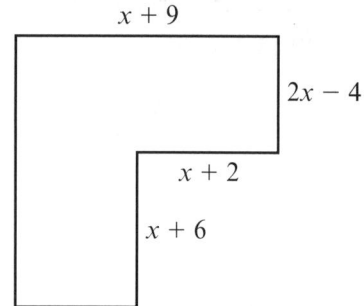

All the measurements are in centimetres.
The perimeter of the shape is 94 cm.
Work out the value of x.
You must show your working.

.... **(5 marks)**

17 Amy shares a bag of sweets with her friends.

She gives Beth $\frac{2}{5}$ of the sweets.

She gives Carl $\frac{3}{10}$ of the sweets.

She has 12 sweets left.

How many sweets does Amy give to Beth?

.... **(4 marks)**

18 Here are the first four terms in an arithmetic sequence:

    7   11   15   19

Is the number 127 a term in this arithmetic sequence?
You must give a reason for your answer.

.... **(3 marks)**

19 A, C and B are three places on a map.
ACB is a straight line.
Construct the perpendicular to the line AB at the point C.
You must leave all your construction lines.

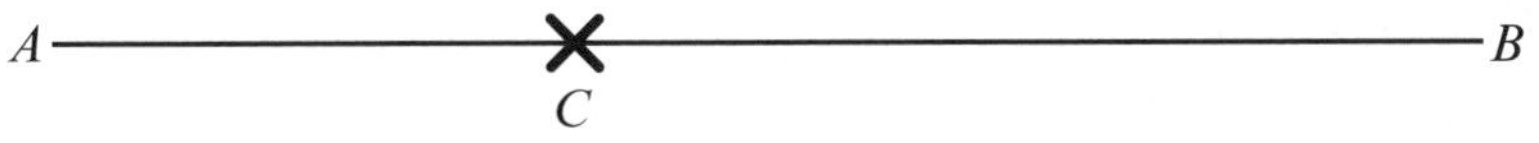

A ————————✕———————————— B
                  C

(2 marks)

20 (a) (i) Write 50 000 in standard form.

....

    (ii) Write 9.6×10^{-5} as an ordinary number.

.... **(2 marks)**

  (b) Work out the value of $(5 \times 10^4) \times (3 \times 10^6)$.
      Give your answer in standard form.

.... **(2 marks)**

21 Shapes *ABCD* and *PQRS* are mathematically similar.

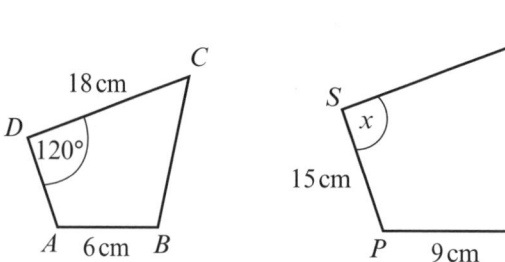

(a) Write down the size of the angle marked *x*.

.... **(1 mark)**

(b) Work out the length of *AD*.

.... **(2 marks)**

(c) Work out the length of *RS*.

.... **(2 marks)**

22 (a) Complete the table of values for $y = \frac{8}{x}$

x	0.5	1	2	4	5	8
y		8			1.6	

(2 marks)

(b) Draw the graph of $y = \frac{8}{x}$

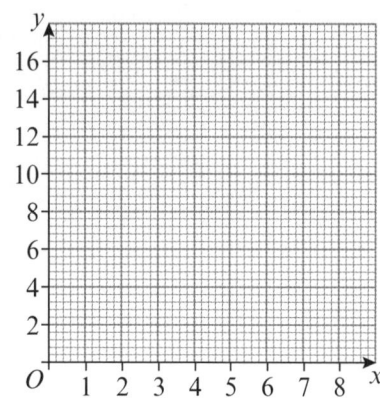

(2 marks)

TOTAL FOR PAPER = 80 MARKS

Practice exam papers for Paper 2 and Paper 3 are available to download free from the Pearson website. Scan this QR code or visit http://activetea.ch/1MpXwRb

Answers

NUMBER

1. Place value
1. (a) 9351
   (b) Four thousand, one hundred and ninety-six
   (c) 5000
2.

Ten thousands	Thousands	Hundreds	Tens	Units
1	2	0	6	0

3. (a) 49, 127, 146, 165, 169
   (b) 7028, 7249, 7429, 7924, 7942
4. (a) £62 400, £63 004, £63 452, £63 593, £65 601
   (b) 36p, 63p, £1.02, £1.12, £1.20
5. He is incorrect.
6. 110 paper plates

2. Negative numbers
1. (a) −11, −4, 0, 4, 6
   (b) (i) −2   (ii) −3   (iii) −10   (iv) −16
2. (i) −14   (ii) −7   (iii) 24   (iv) 7
3. (a) −2°C   (b) −7°C   (c) −15°C   (d) 22°C
4. (a) 50°C
   (b) New Delhi
   (c) No, answer is −9°C

3. Rounding numbers
1. (a) 27 000   (b) 6500   (c) 87 540
2. (a) 9   (b) 8.6   (c) 8.64
3. (a) 0.003   (b) 0.0035   (c) 0.003 47
4. (a) 40 000   (b) 39 000   (c) 38 700
5. (a) 20.4 g   (b) 300 g   (c) 170 g   (d) 130 g
6. No, correct answer is 0.0235 g

4. Adding and subtracting
1. (a) 1023   (b) 577
2. (a) 8178   (b) 177
3. £11.40
4. 41
5. He is not correct. Coffee costs £2.06

5. Multiplying and dividing
1. (a) 1909   (b) 243
2. (a) 112   (b) 196
3. 756
4. (a) 43 290   (b) 34
5. (a) 5 boxes   (b) 9 boxes
6. (a) 125 chocolates   (b) 19 chocolates

6. Decimals and place value
1. (a) 7 tenths or $\frac{7}{10}$ (b) 8 hundredths or $\frac{8}{100}$
   (c) 4 thousandths or $\frac{4}{1000}$
2. 1.4, 3.2, 6.2, 6.4, 12.8
3. 0.05, 0.6, 0.61, 0.611, 0.613
4. 0.7, 0.725, 0.73, 0.778, 0.78
5. (a) 2451   (b) 24.51   (c) 4.3
6. He is incorrect because 435.2 ÷ 13.6 = 32

7. Operations on decimals
1. (a) 14.63   (b) 75.36   (c) 117.12
   (d) 0.0329   (e) 13.9   (f) 63
2. £282.60
3. £15.40
4. £20.94

8. Squares, cubes and roots
1. (a) 16   (b) 8   (c) 9
   (d) 8   (e) 4   (f) 2
   (g) 3   (h) −4   (i) −5
2. (a) 81   (b) 125   (c) 12   (d) 6
3. 52
4. (a) 36 or 49   (b) 8   (c) 49
5. No because $2 \times 2 \times 2 = 8$
6. No because $16 + 4 + 1 = 21$ which is odd

9. Indices
1. (a) 4^2 (b) 4^5
2. (a) 5^9 (b) 5^3 (c) 5^4 (d) 5^{12}
3. (a) 9^{-1} (b) 9^{-4}
4. (a) 3^3 (b) 3^2 (c) 3^8 (d) 3^6
5. (a) 1   (b) $\frac{1}{7}$ (c) $\frac{1}{49}$ (d) $\frac{1}{64}$
   (e) $\frac{27}{64}$ (f) $\frac{25}{16}$
6. $x = 8$

10. Estimation
1. (a) 14 000   (b) 6   (c) 125 000
2. 125
3. 7200
4. 17 500
5. 750
6. 36 000
7. (a) 432 cm²   (b) Underestimate

11. Factors, multiples and primes
1. (a) $1 \times 36, 2 \times 18, 3 \times 12, 4 \times 9, 6 \times 6$
   (b) 7, 14, 21, 28, 35, 42, 49, 56, 63, 70
2. (a) factor   (b) multiple
3. 41, 43, 47
4. (a) 2 or 6   (b) 21 or 49   (c) 6 and 8
5. $14 + 7 + 1$ or $14 + 7 + 2$
6. (a) 2×3^3 (b) $2^5 \times 3$ (c) $2 \times 3^2 \times 7$ (d) $2^2 \times 3^2 \times 7$

12. HCF and LCM
1. (a) 12   (b) 60
2. (a) (i) $2 \times 3^2 \times 5$ (ii) $2 \times 3 \times 5 \times 7$
   (b) 30   (c) 630
3. (a) 12   (b) 144

13. Fractions
1.
2. (a) $\frac{1}{2}$ (b) $\frac{2}{3}$ (c) $\frac{7}{24}$ (d) $\frac{2}{7}$
3. (a) $\frac{3}{4}$ (b) $\frac{3}{4}$
4. (a) £45   (b) £64   (c) £140   (d) £150
5. £17.50
6. £66

14. Operations on fractions
1. (a) $\frac{11}{15}$ (b) $\frac{11}{20}$ (c) $\frac{69}{56}$ (d) $-\frac{1}{63}$
2. (a) $\frac{1}{6}$ (b) $\frac{15}{44}$ (c) $\frac{8}{3}$ (d) $\frac{3}{2}$
3. $\frac{4}{15}$
4. (a) $\frac{8}{35}$ (b) 48 litres

15. Mixed numbers
1. (a) $6\frac{11}{20}$ (b) $2\frac{1}{10}$
2. (a) $3\frac{5}{6}$ (b) $3\frac{1}{3}$
3. (a) 9   (b) $3\frac{9}{13}$
4. $5\frac{11}{12}$ hours
5. $1\frac{1}{18}$ litres
6. 12

16. Calculator and number skills
1. (a) 15   (b) 95   (c) 18   (d) 81
2. (a) 6   (b) 1   (c) 6

3   1.751 592 357
4   (a) 8.5625   (b) 9
5   (a) 1.248 005 424   (b) 1.25
6   (a) 2.798 083 024   (b) 2.8

17. Standard form 1
1   (a) 4.5×10^4 (b) 0.000 034   (c) 2.8×10^7
2   (a) 5.67×10^5 (b) 5.67×10^{-5} (c) 5.67×10^{10}
3   (a) 6.74×10^6 (b) 7.3×10^6 (c) 6.2×10^6
4   (a) 2.05×10^8 (b) 7.5×10^7
5   9.3×10^4 km/h

18. Standard form 2
6   (a) 1.8×10^4 (b) 2×10^{20}
7   (a) 7.01×10^4 (b) 7.52×10^5
8   (a) 1.12×10^{10} (b) 4.48×10^8 (c) 7×10^{10}
9   1.44×10^8 km
10   15 000 s

19. Counting strategies
1   (A, 1), (A,2), (A,3), (R,1), (R,2), (R,3), (T,1), (T,2), (T,3)
2   (C,P), (C,G), (C,B), (L,P), (L,G), (L,B), (V,P), (V,G), (V,B)
3   (W,X), (W,Y), (W,Z), (X,Y), (X,Z), (Y,Z)
4   6
5   6
6   20

20. Problem-solving practice 1
1   2, 5, 13, 17 (there are other possibilities, for example, 2, 3, 11, 19)
2   £7
3   10 cups of coffee
4   (a) 12 600
    (b) She is incorrect, $72 \times 6 = 432$, so she needs 7 vans.
5   $\frac{2}{3}$
6   60 boxes

21. Problem-solving practice 2
7   27
8   6
9   18 cm
10   9 am
11   $(4.86 \times 10^{-5}) \times (6.2 \times 10^4) = 3.01 \approx 3$ m
12   $\frac{14}{5}$

ALGEBRA

22. Collecting like terms
1   (a) expression   (b) formula   (c) equation
2   (a) $5x$ (b) $6xy$
3   (a) $4x + 2y$ (b) $5ab$ (c) $t + 11v$ (d) $3c - 2d$
4   (a) $2x$ (b) $3t^2$ (c) $3a + 8b + 7$ (d) $x - 7y$
5   (a) $3n$ (b) $6p + q$ (c) $3m + 4n$ (d) $2a + 6b + 5$
6   (a) $2y^2$ (b) $3x^2 + 2x$

23. Simplifying expressions
1   (a) y^2 (b) $3mt$
2   (a) w^4 (b) $28d$ (c) $30k$ (d) $40jk$
3   (a) $15x^2$ (b) $6ef$ (c) $4a$ (d) $8b$
4   (a) $35gh$ (b) $8t^3$ (c) $3x$ (d) $4y$
5   (a) $24abc$ (b) $8p$
6   $2(x \times y)$ and $4xy \div 2$

24. Algebraic indices
1   (a) a^9 (b) a^3 (c) a^4 (d) a^{12}
2   (a) t^3 (b) t^3 (c) t^8 (d) t^6
3   (a) x^{12} (b) $64x^6$ (c) $8x^9$ (d) $12x^7$
    (e) $12x^7y^5$ (f) $3x^2y^3$
4   (a) 8   (b) 5   (c) 5
5   1.5

25. Substitution
1   240 km
2   (a) 23   (b) −11

3   (a) 34   (b) 19
4   (a) −6   (b) 100   (c) 12
5   (a) 104   (b) 23   (c) 20
6   Abbie is correct, $\frac{1}{2} \times 2 \times 3^2 = 9$

26. Formulae
1   50 minutes
2   £175
3   30
4   36
5   18
6   52
7   −4°F
8   $y = 3x^2 - 4x + 3$
    $y = 3(-2)^2 - 4(-2) + 3$
    $y = 12 + 8 + 3$
    $y = 23$

27. Writing formulae
1   $4g + 5h$
2   $S = 10m + 20n$
3   $P = 30n + 50$
4   $T = 8(m + 5)$
5   $P = 10x + 2$
6   (a) $B = n + 4$
    (b) Yes, because Carl is $3n$ years old

28. Expanding brackets
1   (a) $3x + 6$ (b) $4x + 20$ (c) $5x - 15$
    (d) $12x + 18$ (e) $x\sqrt{2} - 2$ (f) $21x - 56$
2   (a) $-3x + 9$ (b) $-4x - 12$ (c) $-6x + 30$
    (d) $-4x - 6$ (e) $-8x + 2$ (f) $-2x + 4$
3   (a) $x^2 + x$ (b) $x^2 + 5x$ (c) $2x^2 - 18x$
    (d) $6x^2 - 9x$ (e) $-2x^2 + 3x$ (f) $-12x^2 + 15x$
4   (a) $7x + 6$ (b) $5x + 14$ (c) $13x - 11$
    (d) $6x^2 - 20x$
5   $a = 20, b = -19$
6   $p = 3, q = 18$

29. Factorising
1   (a) $3(x + 2)$ (b) $6(a + 3)$ (c) $2(p - 3)$
    (d) $5(y - 3)$ (e) $3(t + 8)$ (f) $4(g - 5)$
2   (a) $x(x + 6)$ (b) $x(x - 4)$ (c) $x(x - 9)$
    (d) $x(x - 12)$ (e) $x(x + 5)$ (f) $x(x - 1)$
3   (a) $3p(p + 2)$ (b) $8y(y - 3)$ (c) $9t(t - 4)$
4   (a) $4d(d + 3)$ (b) $6x(x - 3)$ (c) $7n(n - 5)$
5   (a) $2x^2 - 3xy, 3, 3x, 2x - 3y$
    (b) $4, (3mn + m^2), 4m, (3n + m), m, (12n + 4m)$
6   $10x + 15 - 4x + 12$
    $= 6x + 27$
    $= 3(2x + 9)$

30. Linear equations 1
1   (a) 16   (b) −5   (c) −15
    (d) −36   (e) −120   (f) −9
2   (a) 5   (b) 12   (c) 2
    (d) 3   (e) 19   (f) −46
3   (a) 6   (b) −3   (c) −4
    (d) −4   (e) 15   (f) −24
4   (a) $5t - 9$ (b) $36 = 5t - 9$ (c) 9

31. Linear equations 2
5   (a) 2   (b) −8   (c) $\frac{3}{4}$
6   (a) 5   (b) 2   (c) −3   (d) 3
7   $\frac{7}{2}$

32. Inequalities
1   (a) $x \leq 4$ (b) $x > -1$
    (c) $-2 < x < 4$ (d) $-1 < x \leq 5$

2 (a)
(b)
(c)
(d)

3 (a) $x = -2, -1, 0, 1, 2, 3, 4$
(b) $x = -2, -1, 0, 1, 2$
(c) $x = -3, -2, -1, 0, 1$

4 (a) $175 \leq a < 185$
(b) $8500 \leq b < 9500$
(c) $123.45 \leq c < 123.55$
(d) $10.75 \leq d < 10.85$

5 Jeff is correct.

33. Solving inequalities

1 (a) $x \leq 10$ (b) $x > 5$ (c) $x \geq 4$
(d) $x \leq -\frac{16}{3}$ (e) $x > 2$ (f) $x \geq -\frac{2}{3}$

2 (a) $x \geq 6$ (b) $x < 7$ (c) $x > 3$
(d) $x \leq -3$

3 (a) $x = -2, -1, 0, 1$
(b) $x = -1, 0, 1, 2, 3, 4$

4 $x = 5$

34. Sequences 1

1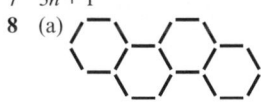

2 (a) 18, 22 (b) 23, 28 (c) 81, 243 (d) 25, 36
3 1, 3, 4, 7, 11, 18
4 (a) 22, 18
(b) Ravina is incorrect. All terms end in even digits.
5 32

35. Sequences 2

6 (a) $4n + 1$ (b) $3n - 1$ (c) $7n - 5$ (d) $5n + 3$
7 $3n + 1$
8 (a)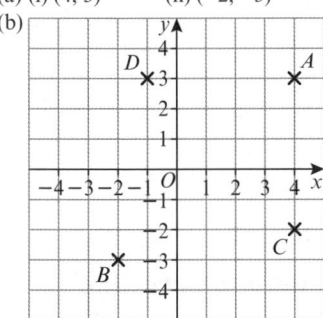
(b) $S = 5n + 1$
9 (a) $4n - 1$
(b) $4n - 1 = 199$
$4n = 200$
$n = 50$
n is an integer therefore 199 is part of the sequence.

36. Coordinates

1 (a) (i) (4, 3) (ii) (−2, −3)
(b)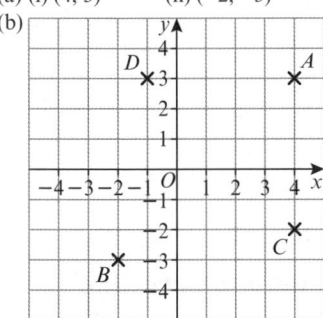

2 (a) (5, 9) (b) $(5, \frac{11}{2})$ (c) (−2, 10) (d) (−4, 3)
3 (a) $(\frac{13}{2}, 10)$ (b) $(\frac{11}{2}, \frac{17}{2})$
4 B (8,6) and then P (2,10)

37. Gradients of lines

1 (a) $\frac{3}{2}$ (b) 2
2 (a) $-\frac{8}{5}$ (b) $-\frac{3}{5}$
3 $\frac{20}{8} = 2.5$ cm/s

38. Straight-line graphs 1

1 (a) −5, −3, −1, 1, 3, 5
(b)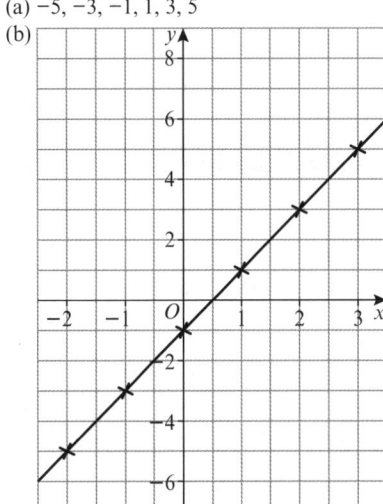

2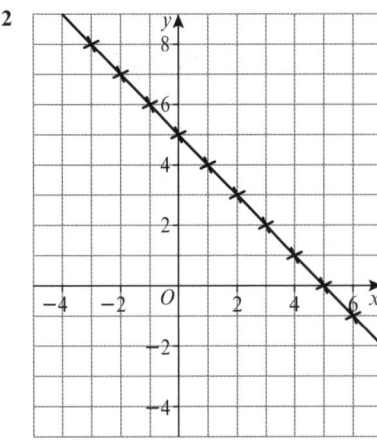

3 $y = 2x + 4$

39. Straight-line graphs 2

4 (a) $y = 3x − 1$ (b) $y = −2x + 12$
(c) $y = 4x + 15$ (d) $y = 4x − 2$
5 (a) $y = 2x − 4$ (b) $y = −3x + 2$
6 $y = 4x + 5$

40. Real-life graphs

1 (a) 96 km (b) Marseille
2 (a) 4.6 m (b) Sandeep
3 (a) £60 (b) Yes, gradient = 1.5

41. Distance–time graphs

1 (a) 10:00 (b) 6 km
(c) 15 minutes (d) 18 km/h
2 (a) 30 minutes (b) 2 km
(c)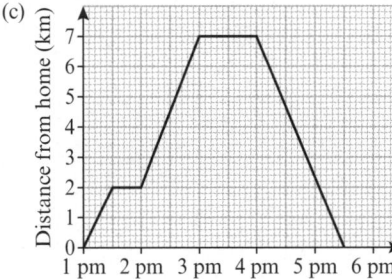

3 (a)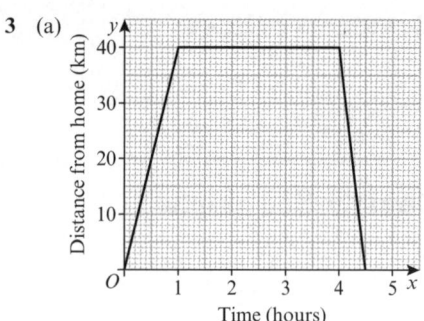
(b) 80 km/h

42. Rates of change
1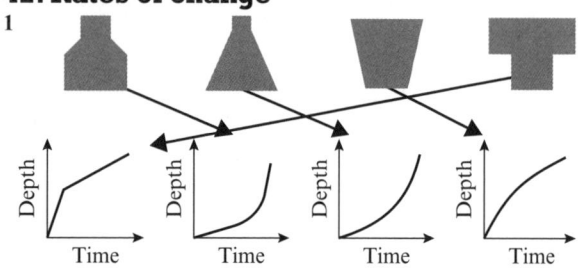
2 (a) £2400 (b) 300
(c) For every month, Dan saves £300.
3 (a) 3 m/s^2 (b) The velocity is constant at 30m/s.
(c) −1.5 m/s^2

43. Expanding double brackets
1 (a) $x^2 + 8x + 15$
(b) $x^2 + 5x + 6$
(c) $x^2 + 5x + 4$
(d) $x^2 − 3x − 10$
(e) $x^2 − 7x + 10$
2 (a) $15x^2 − 23x + 4$
(b) $16x^2 − 14x + 3$
(c) $28x^2 − 55x + 25$
(d) $x^2 + 6x + 9$
(e) $4x^2 − 20x + 25$
3 $n(n + 1) + 1(n + 1) − n^2$
$= n^2 + 2n + 1 − n^2$
$= 2n + 1$ which is odd
4 $x^2 + y^2 − (x^2 − 2xy + y^2)$
$= 2xy$ which is even
5 $(2x + 3)(x − 2)$
$= 2x^2 − 4x + 3x − 6$
$= 2x^2 − x − 6$

44. Quadratic graphs
1 (a)

(b)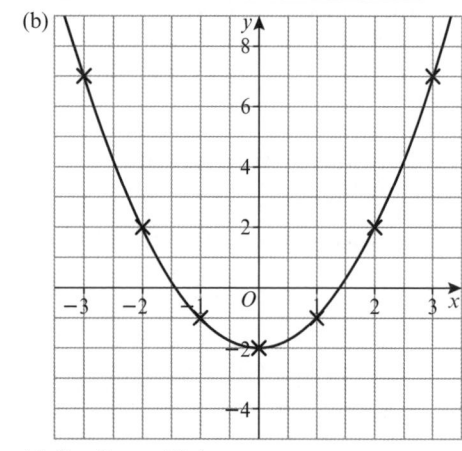

(c) (0, −2) (d) 4

2 (a)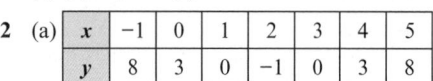

(b)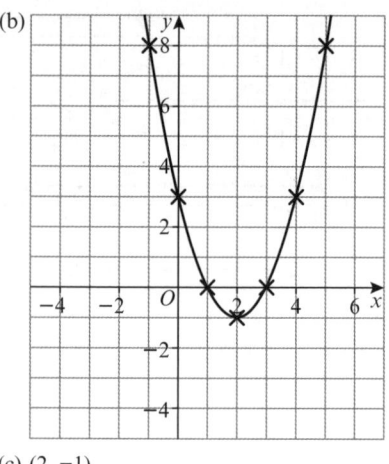

(c) (2, −1)

45. Using quadratic graphs
1 (a)

(b)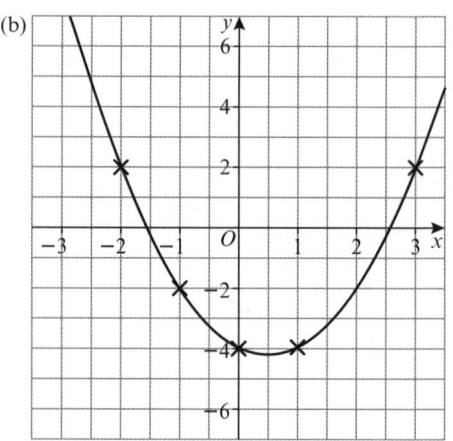

(c) (i) −4.25 (ii) −1.6, 2.6
(d) Graphical method so not accurate

2 (a) | x | −2 | −1 | 0 | 1 | 2 | 3 |
|---|---|---|---|---|---|---|
| y | −6 | 3 | 8 | 9 | 6 | −1 |

(b)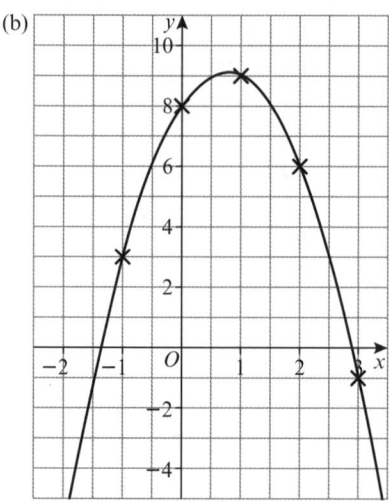

(c) (0.8, 9.1)

46. Factorising quadratics
1 (a) $(x + 1)(x + 3)$ (b) $(x + 10)(x + 1)$
(c) $(x + 5)(x + 1)$ (d) $(x − 10)(x − 1)$
(e) $(x − 2)(x − 10)$ (f) $(x − 7)(x − 2)$
2 (a) $(x + 7)(x − 1)$ (b) $(x + 5)(x − 1)$
(c) $(x − 5)(x + 3)$

3 (a) $(x-11)(x-2)$ (b) $(x-8)(x+2)$
  (c) $(x-4)(x-10)$
4 (a) $(x-3)(x+3)$ (b) $(x-12)(x+12)$
  (c) $(x-9)(x+9)$ (d) $(x-8)(x+8)$
  (e) $(x-1)(x+1)$ (f) $(x-13)(x+13)$

47. Quadratic equations
1 (a) $x=0, x=3$ (b) $x=0, x=-5$
  (c) $x=0, x=7$
2 (a) $x=-4, x=-2$ (b) $x=4, x=3$
  (c) $x=-5, x=-4$ (d) $x=-1, x=-7$
  (e) $x=6, x=-4$
3 (a) $x=-2, x=2$ (b) $x=-5, x=5$
  (c) $x=-7, x=7$ (d) $x=-11, x=11$
  (e) $x=-3, x=3$
4 10 and 11

48. Cubic and reciprocal graphs
1 (a)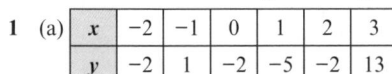

x	-2	-1	0	1	2	3
y	-2	1	-2	-5	-2	13

(b)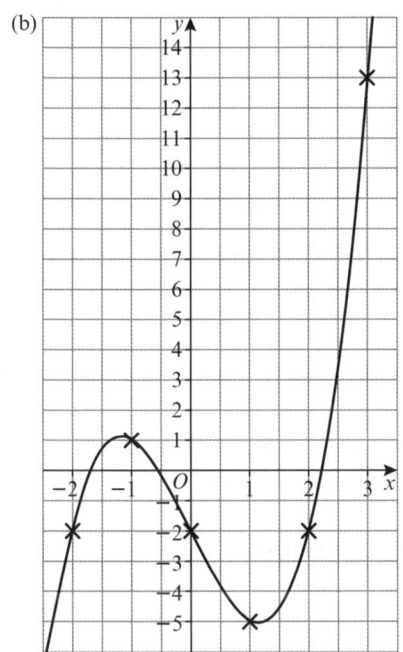

(c) $(-1.2, 1.1), (1.2, -5.1)$
(d) $x=-2.1, x=0.25$ and $x=1.8$
2 (i) B (ii) D (iii) A (iv) C (v) E

49. Simultaneous equations
1 (a) $x=3, y=2$ (b) $x=5, y=-2$
2 (a)
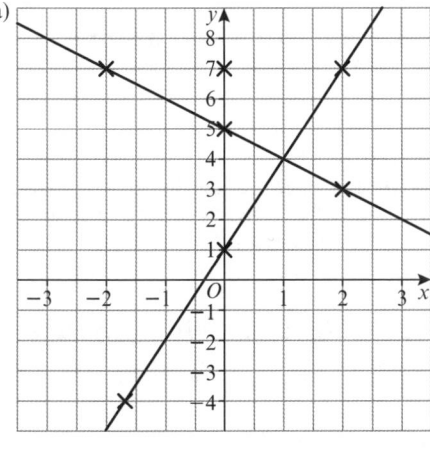

$x=1, y=4$

(b)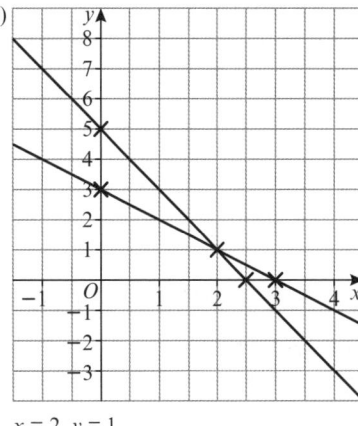

$x=2, y=1$

50. Rearranging formulae
1 $x=3$
2 $t=2.5$ seconds
3 (a) $t=\frac{1}{10}(v-u)$ (b) $n=\frac{1}{6}(m-19)$
  (c) $u=\frac{1}{t}(d-at^2)$ (d) $D=\frac{1}{6}(A-P)$
4 (a) $t=\frac{d}{s}$ (b) $h=\frac{4d^2}{5}$
  (c) $t=\frac{2s}{(u+v)}$ (d) $s=\frac{(v^2-u^2)}{2a}$
5 (a) $n=\frac{P}{h}-2$ (b) $x=\frac{1}{2}-\frac{t}{6}$

51. Using algebra
1 (a) $7x-9=96$ (b) 15
2 15 m
3 15 m and 16 m
4 6

52. Identities and proof
1 $(2n-1)^2 = (2n-1)(2n-1) = 4n^2 - 4n + 1$
  $(2n+1)^2 = (2n+1)(2n+1) = 4n^2 + 4n + 1$
  $(2n-1)^2 + (2n+1)^2 = 4n^2 - 4n + 1 + 4n^2 + 4n + 1$
  $= 8n^2 + 2 = 2(4n^2 + 1)$
2 $n + (n+1) + (n+2) + (n+3) = 4n + 6 = 2(2n+3)$ therefore a multiple of 2
3 $5(x-c) = 4x - 5$
  $5x - 5c = 4x - 5$
  $x = 5c - 5$
  $x = 5(c-5)$ therefore x is a multiple of 5
4 (a) $(x-1)^2 \equiv (x-1)(x-1) \equiv x^2 - 2x + 1$
  (b) $(x+1)^2 \equiv x^2 + 2x + 1$
  $(x+1)^2 + (x-1)^2 \equiv x^2 + 2x + 1 + (x^2 - 2x + 1)$
  $\equiv 2x^2 + 2 \equiv 2(x^2 + 1)$
5 (a) $2n + 2n + 2 + 2n + 4 = 6n + 6 = 6(n+1)$ therefore always a multiple of 6
  (b) 2, 6, 8

53. Problem-solving practice 1
1 8 days
2 (a) (6, 1) (b) (4, 3)
3 (a) £130 (b) 6 hours
4 Dan is correct.
  $4 \times 2 + 5 \times 2^2 = 8 + 5 \times 4 = 8 + 20 = 28$
5 nth term is $4n+2$
  10th term is $4 \times 10 + 2 = 42$
  She is not correct.

54. Problem-solving practice 2
6 35 cm
7 55°
8 (a) $y = 5x + k$ where k is any number
  (b) $y = 3x - 5$

147

9   nth term is $3n - 1$
     $3n - 1 = 34$
     $3n = 35$
     $n = \frac{35}{3}$
     n is not an integer therefore 34 is not a term in this linear sequence.
10   $(3x + 4)(2x - 1) = A$
     $6x^2 - 4 + 8x - 3x = A$
     $6x^2 + 5x - 4 = A$

RATIO & PROPORTION

55. Percentages
1   (a) 4.5   (b) 11
2   (a) 62.5%   (b) 20%
3   (a) £7.68   (b) £103.68
4   £13 200
5   37.5%
6   (a) 30.8%
     (b) $\frac{30}{100} \times 140 = 42$ (French)
     $\frac{60}{100} \times 180 = 108$ (German)
     $\frac{150}{320} \times 100 = 46.9\% = 47\%$ (to 2 s.f.)

56. Fractions, decimals and percentages
1   (a) $\frac{6}{25}$ (b) $\frac{16}{25}$ (c) $\frac{18}{25}$
2   (a) $\frac{3}{10}$, 61%, 0.62   (b) 0.32, 33%, $\frac{7}{20}$
     (c) 37%, 0.38, $\frac{2}{5}$
3   £575
4   48
5   $\frac{30}{100} \times £2100 = £630$ (Amy)
     $\frac{1}{3} \times £1800 = £600$ (Bhavna)
     Amy saves the most money each month

57. Percentage change 1
1   (a) 74.88   (b) 131.04   (c) 84.48   (d) 275.88
2   £126.90
3   (a) 20%   (b) 30%   (c) 35%   (d) 10%
4   (a) 34.8%   (b) 28%
5   £23 320

58. Percentage change 2
6   Kelly-air
7   Postland
8   Footworld

59. Ratio 1
1   (a) 3:2   (b) 27:8   (c) 7:8
2   (a) £20:£30   (b) £100:£250:£400
3   Cheese = 16 g, peppers = 24 g
4   Paul = 48 miles, Faye = 60 miles
5   (a) 1 part is 24, 24 × 3 = £72
     (b) (3 × 24) + (4 × 24) + (9 × 24) = £384

60. Ratio 2
6   (a) 28 g   (b) 40 g
7   £4400
8   48 ÷ 12 = 4
     Flour = 20, margarine = 16 and sugar = 12
     Flour = 1825 ÷ 20 = 91.25, margarine = 700 ÷ 16 = 43.75 and sugar = 250 ÷ 12 = 20.83
     Therefore, maximum number of cakes = 20

61. Metric units
1   (a) 4.5 cm   (b) 720 mm   (c) 3500 m
     (d) 5300 g   (e) 4300 ml   (f) 0.48 g
2   (a) 150 mm   (b) 2.8 cm   (c) 1.8 kg
     (d) 2.8 km   (e) 0.053 litres   (f) 145 000 mg
3   16
4   80
5   No, he can only fit 19.

62. Reverse percentages
1   £60
2   £33 250
3   £33 600
4   No, Kate earned £518.52 last year.
5   Alison invested £1650 and Nav invested £1680. Nav invested more than Alison.

63. Growth and decay
1   £17 569.20
2   $n = 3$
3   (a) £12 528.15   (b) £6332.78
4   (a) 6.5%   (b) £2815.71
5   It is worth £110.74

64. Speed
1   8.9 m/s
2   425 km/h
3   3 hours 20 minutes
4   240 km
5   35 ÷ 0.25 = 140 km/h
     140 is greater than 130
6   Karen had the lower average speed.
7   100 m race

65. Density
1   0.875 g/cm^3
2   147 g
3   10 375 cm^3
4   432 g
5   5666.4 g
6   Gavin is not correct, it is bronze.

66. Other compound measures
1   400 N/m^2
2   18 750 N/m^2
3   0.006 25 m^2
4   35 minutes
5   12 cm × 100 cm × 100 cm = 120 000 cm^3
     120 000 ÷ 2 = 60 000
     60 000 cm^3 = 60 000 ml
     60 000 ÷ 250 = 240 s
     240 seconds is 4 minutes

67. Proportion
1   £2.80
2   £8.25
3   32
4   20 days
5   8 days
6   £16
7
Large	Medium
£:g	£:g
4.80:200	4.50:175
0.024:1	0.026:1

The large basket is better value for money.

68. Proportion and graphs
1   1170
2   7.5
3   $x = \frac{1}{2y}$
4   (a) 5 N
     (b) The graph is a straight line passing through the origin / there is a constant increase / as extension increases, force increases.
5   (a) 2 litres   (b) As pressure increases, volume decreases.

69. Problem-solving practice 1
1   Statistics
2   False, (122 ÷ 500) × 100 = 24.4%
3   Nile is cheaper.
4   5400 ÷ 900 = 6
     $\frac{1}{6}$ of $\frac{1}{5}$ of 900 = 30 ml

70. Problem-solving practice 2
5 No, he needs four more people.
6 1000 g
7 No. She needs (60 ÷ 4) × 15 = 225 g of almonds and she only has 200 g.
8 $\dfrac{100 + 6}{100} = 1.06$
   £14 000 × (1.06)^4 = £17 674.68
   Kim has enough money.

GEOMETRY & MEASURES

71. Symmetry
1 (a) 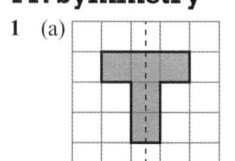   (b)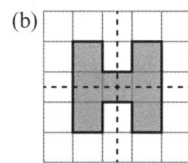
2 (a) 4   (b) 2 or 3   (c) 1
3 (a) 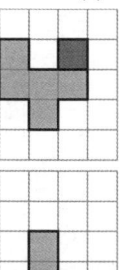   (b)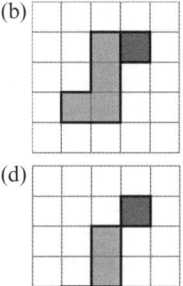
   (c)   (d)
4 (a) 6
   (b) 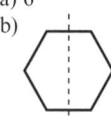 Other possible lines go through opposite corners.

72. Quadrilaterals
1 (a) Rectangle   (b) Trapezium
   (c) Parallelogram (d) Square
   (e) Rhombus   (f) Kite
2 (a) 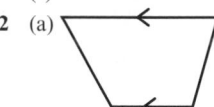   (b)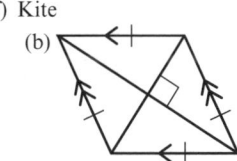
   (c) 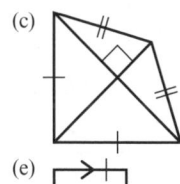 (d)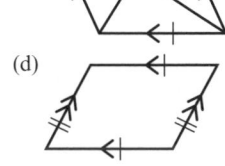
   (e) 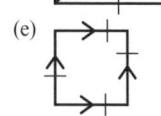 (f)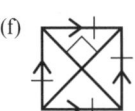

73. Angles 1
1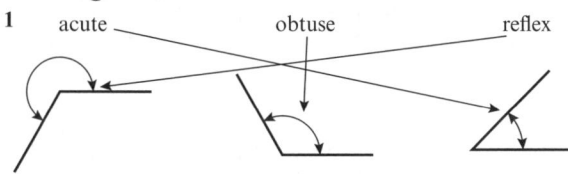
2 (a) (i) Obtuse angle
       (ii) x is more that 90° but less than 180°.
   (b) (i) Reflex angle
       (ii) x is more than 180° but less than 360°.
3 (a) (i) 109°
       (ii) Angles on a straight line add up to 180°.
   (b) (i) 146°
       (ii) Angles around a point add up to 360°.

74. Angles 2
4 292°
5 (a) (i) 60°
       (ii) The triangle is an equilateral triangle.
   (b) 150°
6 (a) 65° because angles on a straight line add up to 180°.
   (b) 65° because x and y are alternate angles.
   (c) 65° because x and z are corresponding angles.

75. Solving angle problems
1 (a) (i) 42°
       (ii) Alternate angles are equal.
       (iii) 111°
   (b) (i) 110°
       (ii) Corresponding angles are equal.
       (iii) 40°
       (iv) Isosceles triangle
2 39°

76. Angles in polygons
1 (a) 72°   (b) 60°   (c) 45°
2 (a) 40°   (b) 9
3 (a) 12   (b) 10   (c) 20
4 135°
5 360° ÷ 6 = 60°
   180° − 60° = 120°
   180° − 120° = 60°
   60° ÷ 2 = 30°

77. Time and timetables
1 (a) 15:15   (b) 02:25   (c) 23:48
2 (a) 4.25 am   (b) 12.10 pm   (c) 8.32 pm
3 52 minutes
4 15:30
5 (a) E   (b) 111 minutes
   (c) 07:45   (d) 09:32

78. Reading scales
1 (a) 23   (b) 340   (c) 5300   (d) 4.6
2 (a)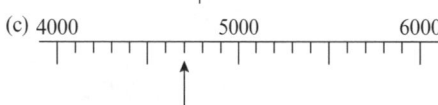
   (b)
   (c)
   (d)
3 (a) 65 km/h
   (b)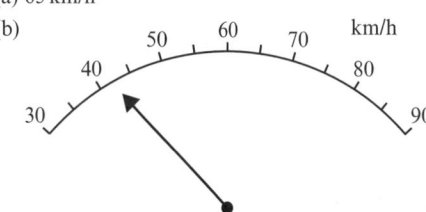
4 0.75 kg

79. Perimeter and area
1 (a) (i) 5 cm^2 (ii) 12 cm
2 (a) 9 cm^2 (b) 6 cm^2
3 (a) 40 cm   (b) 27 cm   (c) 72 cm   (d) 58 cm

80. Area formulae
1. (a) 90 cm² (b) 36 cm² (c) 28 cm² (d) 190 cm²
2. (a) 72 cm² (b) 392 cm²
3. $h = 5$ cm

81. Solving area problems
1. 115 cm²
2. No, the local developer needs to offer £108.75 more.
3. No, she needs 18 tins to paint the wall.

82. 3-D shapes
1. (a) Cube (b) Cuboid (c) Cylinder
   (d) Triangular prism (e) Pyramid (f) Sphere
2. (a) 150 cm² (b) 122 cm² (c) 184 cm²
3.

3-D shape	Number of faces	Number of edges	Number of vertices
(a) Cube	6	12	8
(b) Cuboid	6	12	8
(c) Triangular prism	5	9	6
(d) Tetrahedron	4	6	4

4. (a) 96 cm² (b) No, she needs 11 tins.

83. Volumes of cuboids
1. (a) 125 cm³ (b) 360 cm³ (c) 1344 cm³
2. 6 cm
3. 140 cubes
4. 11.67 cm

84. Prisms
1. (a) 48 cm³ (b) 144 cm³ (c) 384 cm³
2. (a) 108 cm² (b) 216 cm² (c) 366 cm²
3. 352 cm³

85. Units of area and volume
1. (a) 60 000 cm² (b) 1500 mm² (c) 4 000 000 m²
   (d) 50 m² (e) 600 cm² (f) 0.8 km²
2. (a) 22 000 000 cm³ (b) 28 000 mm³
   (c) 3 000 000 000 m³ (d) 200 m³
   (e) 50 000 cm³ (f) 0.42 km³
3. (a) 200 litres (b) 8000 litres (c) 12 000 litres
4. (a) 45 litres (b) 96 000 litres

86. Translations
1. (a) $\begin{pmatrix} 3 \\ 4 \end{pmatrix}$ (b) $\begin{pmatrix} -2 \\ -3 \end{pmatrix}$ (c) $\begin{pmatrix} -5 \\ 6 \end{pmatrix}$
2. (a) and (b)

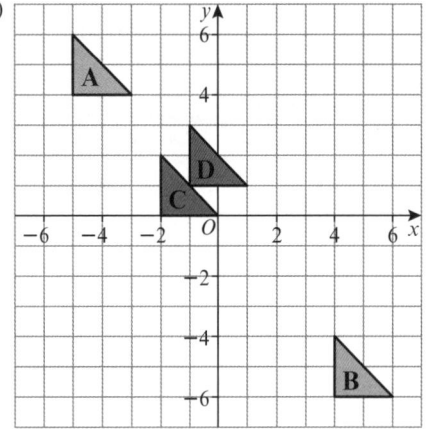

3. (a) translation $\begin{pmatrix} -7 \\ -3 \end{pmatrix}$ (b) translation $\begin{pmatrix} 4 \\ 5 \end{pmatrix}$

87. Reflections
1. (a)

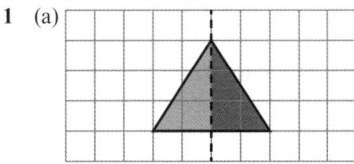

(b)

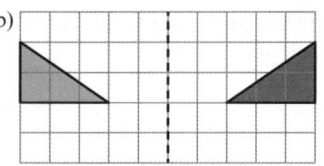

(c)

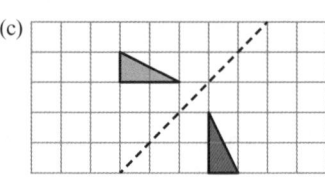

2. (a), (b) and (c)

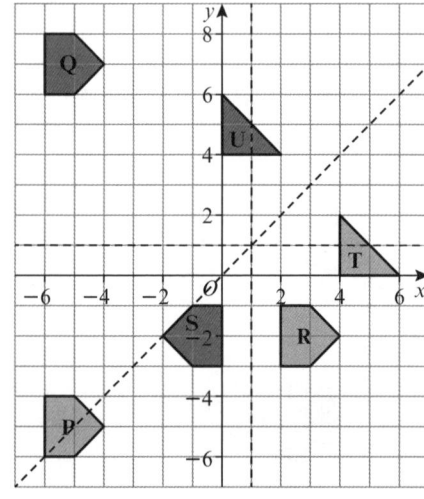

3. (a) Reflection in the line $y = -1$
   (b) Reflection in the line $y = x$

88. Rotations
1. (a) and (b)

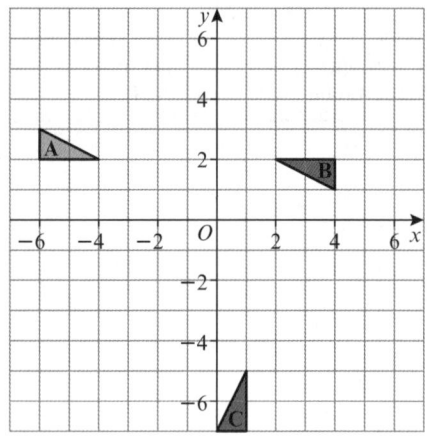

2. (a) and (b)

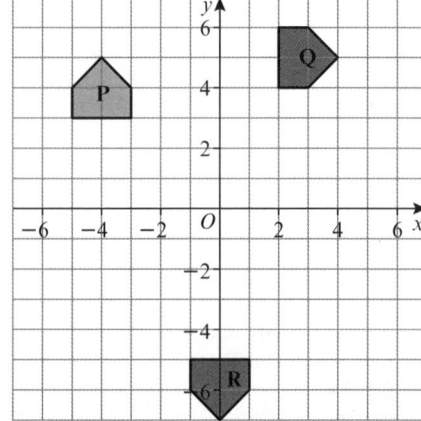

3 (a) Rotation 90° about (1, −1) clockwise
   (b) Rotation 180° about (0, −1)

89. Enlargements
1 (a) 5
   (b)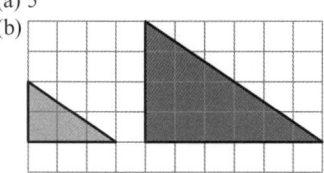

2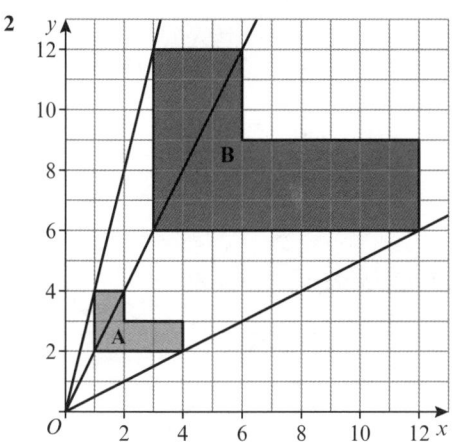

3 (a) Enlargement of scale factor $\frac{1}{3}$ at (4, −4)
   (b)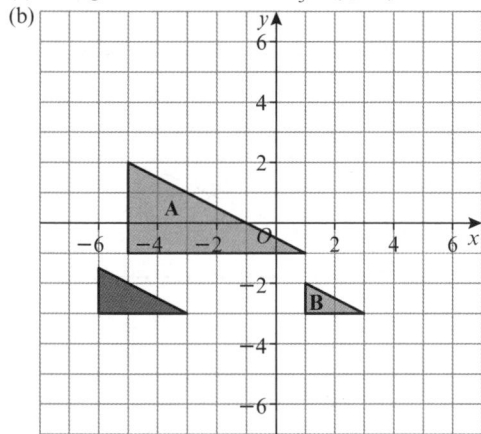

90. Pythagoras' theorem
1 (a) 10.7 cm   (b) 6.75 cm   (c) 15.7 cm
2 1347 cm
3 She is incorrect as the diagonal of the suitcase is less than 125 cm.
4 No, it cannot be totally immersed as the maximum length enclosed by the pool is less than 6 m.

91. Line segments
1 9.22
2 7.07
3 10
4 (a) (1, −2)
   (b) diameter = $\sqrt{8^2 + 6^2} = 10$
       hence, radius = 10 ÷ 2 = 5

92. Trigonometry 1
1 (a) 54.3°   (b) 57.8°
2 62.8°
3 53.0°
4 She cannot use smooth tiles on her roof as angle x is greater than 17° (20.4°).

93. Trigonometry 2
5 (a) 14.3 cm   (b) 16.3 cm
6 5.35 m
7 55.1 m
8 (a) 21.0 m   (b) 60.3°

94. Exact trigonometry values
1
	0°	30°	45°	60°	90°
sin	0	$\frac{1}{2}$	$\frac{1}{\sqrt{2}}$	$\frac{\sqrt{3}}{2}$	1
cos	1	$\frac{\sqrt{3}}{2}$	$\frac{1}{\sqrt{2}}$	$\frac{1}{2}$	0
tan	0	$\frac{1}{\sqrt{3}}$	1	$\sqrt{3}$	–

2 $x = 9$ cm
3 (a) 30°   (b) 30°
4 $30\sqrt{3}$ feet

95. Measuring and drawing angles
1 (a) 45°   (b) 139°   (c) 225°
2 (a)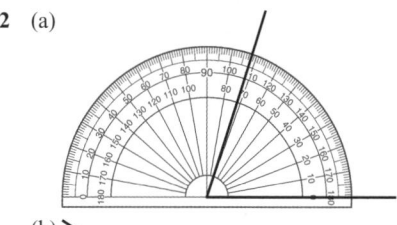
   (b)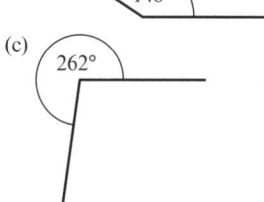
   (c)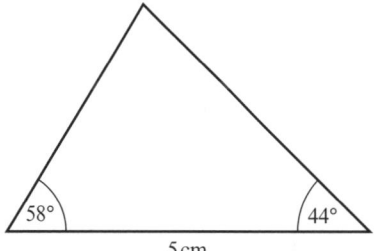

3 (a) 56°, acute   (b) 150°, obtuse   (c) 235°, reflex
4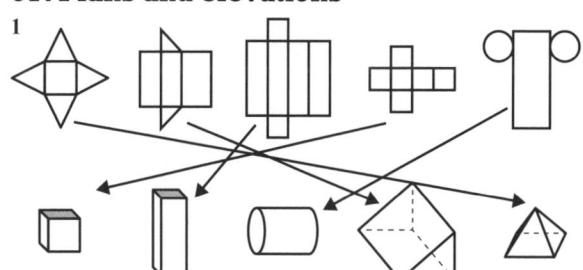

96. Measuring lines
1 (a) 3.2 cm   (b) 4.3 cm   (c) 6 cm
2 (a) A line 52 mm long (±1 mm)
   (b) A line 6 cm long (±1 mm)
   (c) A line 7.8 cm long (±1 mm)
3 Cross marked halfway between A and B (±1 mm)
4 6.7 m
5 (a) 2 m   (b) 6.6 m

97. Plans and elevations
1

2 (a)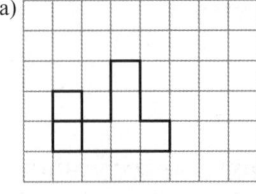

(b)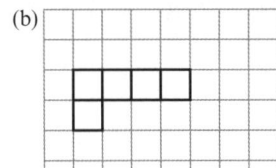

3

4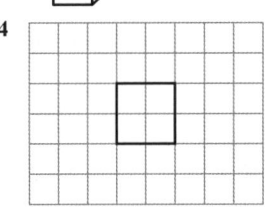

98. Scale drawings and maps
1 (a) 37 m   (b) 22.5 km   (c) 78 km
2 (a) 2.5 km   (b) 12 km   (c) 154 km
3 (a) 20 cm   (b) 15 cm   (c) 5 cm
4 (a) 15 m   (b) 15 cm

99. Constructions 1
1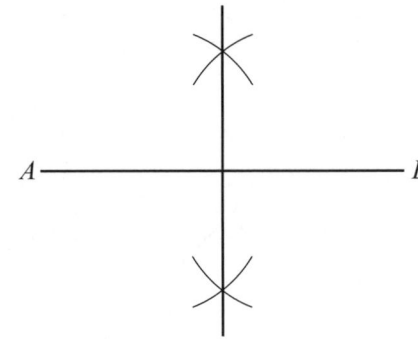

2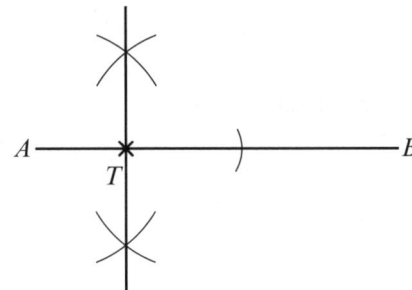

3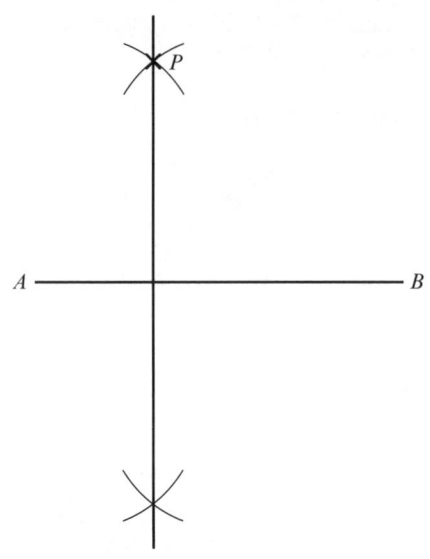

100. Constructions 2
4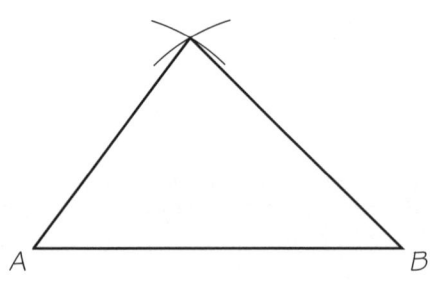

5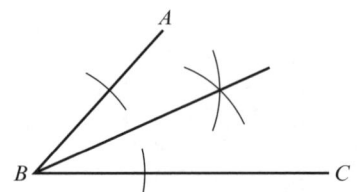

6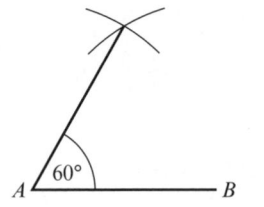

7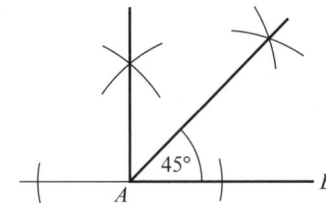

101. Loci
1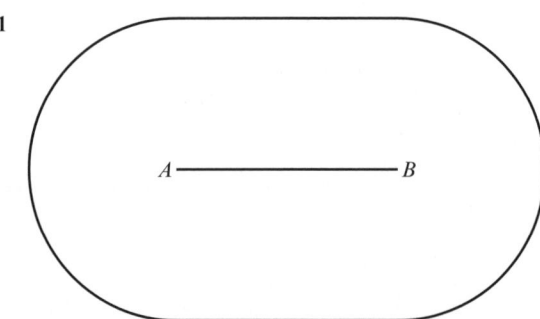

2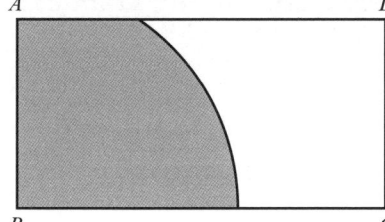

3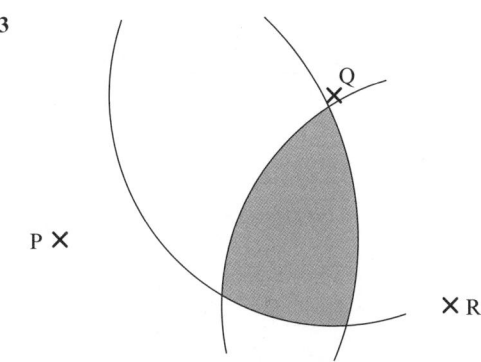

4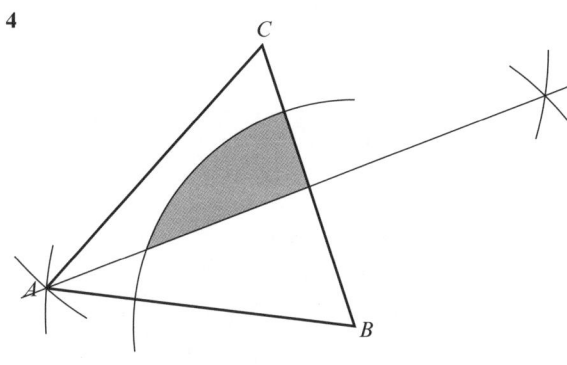

102. Bearings
1. (a) (i) 30°   (ii) 210°
   (b) (i) 130°   (ii) 310°
2. (a)

   (b)

   (c)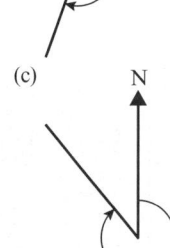

3 (a) 16 km   (b) 25°
   (c)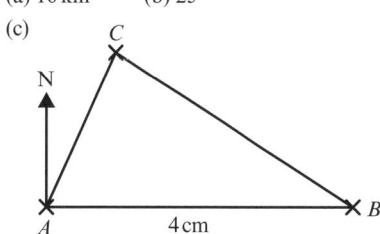

103. Circles
1. (a) chord   (b) radius   (c) diameter
2. (a) 37.7 cm   (b) 50.3 cm
3. (a) 5.57 cm   (b) 14.6 cm
4. (a) 42.8 cm   (b) 72.0 cm
5. 3140 cm

104. Area of a circle
1. (a) 113 cm²   (b) 201 cm²   (c) 531 cm²
2. (a) 113 cm²   (b) 308 cm²   (c) 1360 cm²
3. (a) 204 cm²   (b) 42.1 cm²   (c) 503 cm²
4. Area A $= \frac{1}{4} \times \pi \times x^2 = \frac{\pi x^2}{4}$

   Area B $= 2 \times \frac{1}{2}\pi \times \left(\frac{x}{2}\right)^2 = \frac{\pi x^2}{4}$

105. Sectors of circles
1. (a) 5.59 cm   (b) 34.0 cm   (c) 27.2 cm
2. (a) 24.4 cm   (b) 58.5 cm   (c) 50.2 cm
3. (a) 36.3 cm²   (b) 214 cm²   (c) 145 cm²
4. $\frac{1}{4} \times \pi \times 12^2 - \frac{1}{2} \times 12 \times 12 = 41.09 \approx 41$

106. Cylinders
1. (a) 704 cm³   (b) 11 300 cm³   (c) 25 400 cm³
2. (a) 452 cm²   (b) 2790 cm²   (c) 4810 cm²
3. (a) cylinder = $\pi \times 15^2 \times 18 = 12\,723$ cm³
   cube = $24 \times 24 \times 24 = 13\,824$ cm³
   Volume of the cube is greater.
   (b) cylinder = $(2 \times \pi \times 15 \times 18) + (2 \times \pi \times 15^2) = 3110.2$ cm²
   cube = $24 \times 24 \times 6 = 3456$ cm²
   The cube has the greater surface area.

107. Volumes of 3-D shapes
1. (a) 251 cm³   (b) 7240 cm³   (c) 142 cm³
2. (a) 302 cm³   (b) 905 m³   (c) 300 cm³
3. Volume of cylinder = 240π
   Volume of cone = 120π
4. $\pi \times x^2 \times 9 = \frac{4}{3} \times \pi \times 27$. Rearranging gives $x = 2$

108. Surface area
1. (a) 251 cm²   (b) 1810 cm²   (c) 462 cm²
2. (a) 226 cm²   (b) 650 cm²   (c) 320 cm²
3. 427 cm²

109. Similarity and congruence
1. (a) A and B   (b) D and E

2 (a)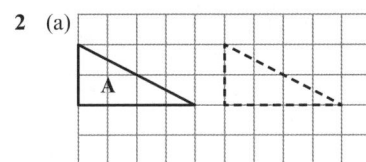

(b)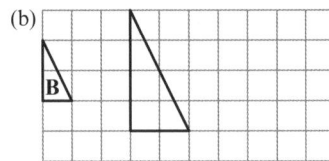

3 (a)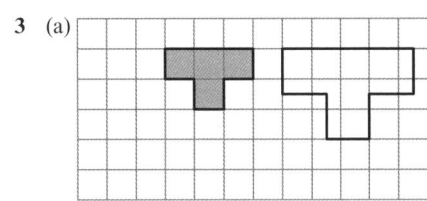

(b)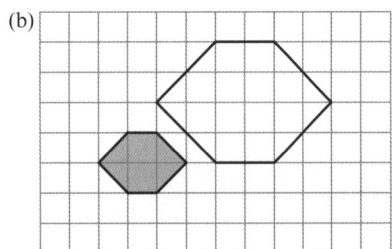

4 Ravina is correct. A and C are the same shape and size. B is an enlargement of A.
5 27

110. Similar shapes
1 (a) 130° (b) 30 cm (c) 18 cm
2 (a) 33 cm (b) 8 cm
3 (a) 6.4 cm (b) 5.7 cm

111. Congruent triangles
1 Side AC = side DF, side AB = DE, side BC = EF therefore SSS
2 $BC = QR$, $CA = PR$, angle BCA = angle PRQ therefore SAS
3 Might be congruent: ABC shows SAS and PQR shows ASA so not definitely congruent, but one angle and one side equal so might be congruent.

112. Vectors
1 (a) $\binom{2}{5}$ (b) $\binom{-2}{-5}$ (c) $\binom{5}{3}$
  (d) $\binom{-5}{-3}$ (e) $\binom{5}{-6}$ (f) $\binom{-5}{6}$
2 (a) **a + b** (b) **−b − a**
3 (a) **p + q** (b) **−q − p** (c) **p − q** (d) **q − p**
4 (a) **a + b** (b) **−a − b**

113. Problem-solving practice 1
1 18°
2 tray = 60 cm × 40 cm × 2 cm = 4800 cm³
  cylinder = $\pi \times 9^2 \times 20$ = 5089 cm³
  There will be no water left in the rectangular tray.
3 Reflection in the line $y = x$

114. Problem-solving practice 2
4 Yes, she does have enough bags.
5 60°
6 (a) 62.0°
  (b) $\sin 42° = \frac{12.8}{BD}$ therefore BD = 19.1 m
  It is long enough.

PROBABILITY & STATISTICS

115. Two-way tables
1
	Bath	Warwick	Lichfield	Total
Boys	10	14	8	32
Girls	7	11	20	38
Total	17	25	28	70

2 (a)
	Dodgeball	Football	Rounders	Total
Girls	12	18	11	41
Boys	6	19	14	39
Total	18	37	25	80

(b) 19 (c) 41 (d) 12

3 (a)
	White	Blue	Red	Total
Motorbikes	7	9	6	22
Cars	3	8	17	28
Total	10	17	23	50

(b) 22 (c) 28 (d) 20

116. Pictograms
1 (a) 8 hours (b) 3 hours
(c)
Monday	○ ○ ○ ○
Tuesday	○ ○ ○
Wednesday	○ ○ ◗
Thursday	○ ○
Friday	○ ◗

2 (a) 20 packets (b) 45 packets
(c)
Monday	▭ ▭ ▭ ▭
Tuesday	▭ ▭ ▫
Wednesday	▭ ▭ ▫
Thursday	▭ ▭ ▭ ▭
Friday	▭ ▭ ▭

117. Bar charts
1 (a) [bar chart: Alsatian 12, Bulldog 18, Labrador 8, Poodle 11]
(b) Bulldog (c) 49

2 (a) Dal (b) 4
(c)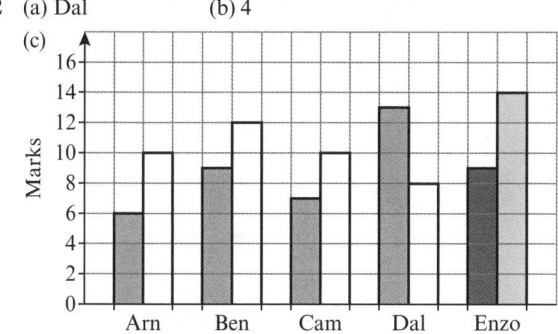
3 1. The scale on the *y*-axis is not linear.
   2. One of the bars is not labelled.

118. Pie charts
1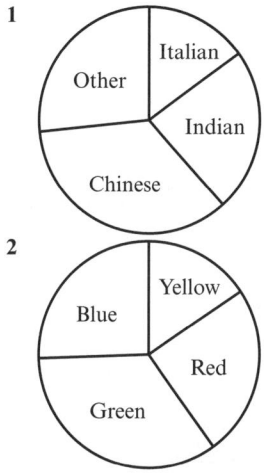
2
3 (a) 10 (b) 20 + 10 + 30 + 60 = 120

119. Scatter graphs
1 (a) Positive (b) 135 g (c) 225 g (d) Yes
  (e) The reading is not within the range of data and we are having to extend the line of best fit.
2 (a) Negative (b) £1250
  (c) This reading is reliable. (d) No

120. Averages and range
1 (a) 7 (b) 10 (c) 9.6 (d) 12
2 (a) 161 cm (b) 160.5 cm
3 (a) 6, 6, 9 (b) 7, 7, 8, 11, 12

121. Averages from tables 1
1 (a) 1 (b) 1 (c) 1.6 (d) 4
2 (a) 5 (b) 5 (c) 4.12
3 (a) The maximum number of chocolates eaten is 5.
  (b) Mean = 2.48
  (c) It will decrease as the value is lower than the mean.

122. Averages from tables 2
4 (a) $8 \leqslant h < 10$ (b) $6 \leqslant h < 8$ (c) $\frac{195}{35} = 5.57$
  (d) Because we are taking the mid-point.
5 (a) The maximum number of minutes is 40.
  (b) 21 minutes
  (c) Yes because 32 minutes is greater than the mean

123. Line graphs
1 (a) 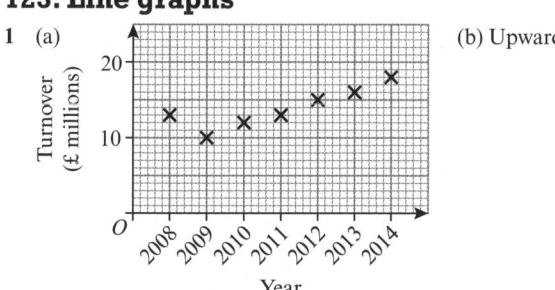 (b) Upwards

2 (a) 4 (b) 60
3 (a) 6 (b) 6.2

124. Stem-and-leaf diagrams
1 (a)
4	7 7
5	0 1 9
6	1 2 3 5 7
7	3 6 7 9
8	6

Key 4|7 means 47 kg
(b) 47 kg (c) 63 kg (d) 86 − 47 = 39

2 (a)
5	3 9
6	1 5 5 5
7	2 5 8 9
8	1 2 3 7 8
9	1 3 4 7

Key 5|3 means 53 beats
(b) 65 beats (c) 79 beats (d) 44 beats
3 (a) No value is repeated and each value only occurs once
  (b) 76 cm (c) 6 (d) the tallest plant

125. Sampling
1 (a) It is quick, cheap and easier than asking the whole class.
  (b) 5.14
  (c) Not very reliable as sample is small
  (d) Ask children in different classes in different years
2 (a) 21.5 cm
  (b) 1.8 m
  (c) Part (a) is more reliable as it is within the data range.
  (d) Carry out more experiments

126. Comparing data
1 (a) 1. Students did better in maths because the mean was higher.
      2. Students' results in maths were more consistent because the range was smaller.
  (b) The amount of rainfall was higher in Dundee because the mean was higher. Wolverhampton's amount of rainfall was more varied because the range was higher.
2 (a) 3.25 (b) 8
  (c) Mr Jones's class had fewer absences, but these were spread over a wider range.
3 The median in class 11A is less than the median in class 11B. The range for both classes is the same.

127. Probability 1
1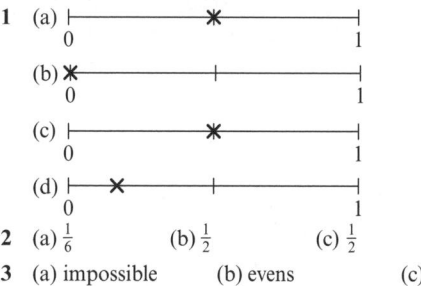
2 (a) $\frac{1}{6}$ (b) $\frac{1}{2}$ (c) $\frac{1}{2}$ (d) 0
3 (a) impossible (b) evens (c) certain
4 (a) $\frac{1}{2}$ (b) $\frac{2}{7}$

128. Probability 2
5 (a) 0.7 (b) 0.3
6 0.21
7 (a) 0 (b) 0.21
8 (a) 0.68 (b) 0.08

129. Relative frequency
1. (a) $\frac{53}{302}$ (b) $\frac{140}{302}$
2. (a) $\frac{1}{5}$ (b) $\frac{17}{50}$ (c) $\frac{41}{50}$
3. (a) $\frac{143}{202}$ = 0.71
   (b) The sample is large so the estimate is accurate

130. Frequency and outcomes
1. (C, P) (C, G) (C, B) (L, P) (L, G) (L, B) (V, P) (V, G) (V, B)
   Probability = $\frac{1}{9}$
2. (a) $\frac{1}{3}$
   (b)
   | Neil's card | X | X | X | Y | Y | Y | Z | Z |
   | Tej's card | X | Y | Z | X | Y | Z | X | Y | Z |

   (c) $\frac{1}{3}$ (d) $\frac{2}{3}$
3. (a)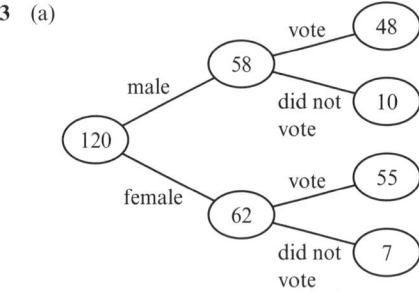
   (b) $\frac{10}{58} = \frac{5}{29}$
4. 35

131. Venn diagrams
1. (a) (i) $x = 15$ (ii) $x = 9$ (iii) $x = 9$
   (b) (i) Students who only study maths
       (ii) Students who don't study French or German
       (iii) Students who study both DT and ICT
2. (a) $\frac{1}{8}$ (b) $\frac{11}{40}$ (c) $\frac{21}{40}$
3. (a) 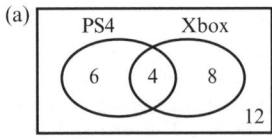 (b) $\frac{12}{30}$ (c) $\frac{14}{30}$

132. Set notation
1. (a) c, e
   (b) m, e, t, r, i, c, g, s
2. 200 is not a member of the universal set.
3. (a)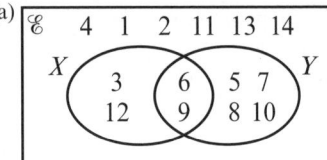
   (b) $\frac{8}{14}$
4.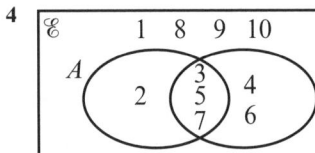
5. (a)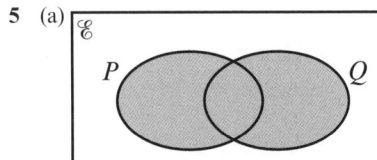

(b)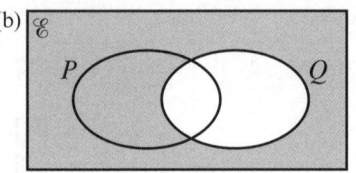

(c)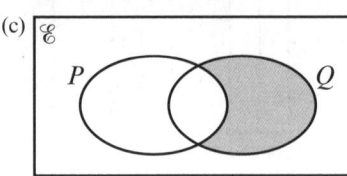

133. Independent events
1. (a) $\frac{9}{100}$ (b) $\frac{49}{100}$ (c) $\frac{42}{100}$
2. (a)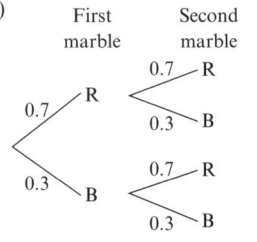
   (b) 0.42
3. (a) 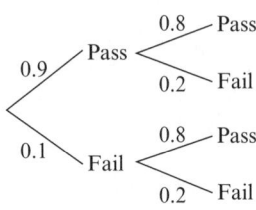 (b) 0.26

134. Problem-solving practice 1
1. (a)
	French	German	Spanish	Total
Female	15	11	13	39
Male	16	17	8	41
Total	31	28	21	80

   (b) $\frac{31}{80}$
2. (a) (i) 0.75 (ii) 0.2 (b) 30
3. The median height of plants in Park A is greater the median height in Park B.
   The range of Park B plant heights is greater than Park A.

135. Problem-solving practice 2
4.
Favourite snack in year 11	Frequency	Angle
Burger	40	80°
Chips	90	180°
Hot dog	20	40°
Kebab	30	60°
Total		180

5. (a)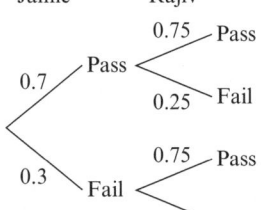
   (b) 0.4

6 (a)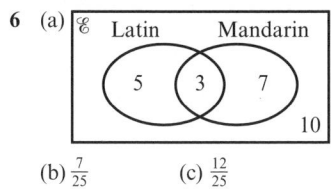

(b) $\frac{7}{25}$ (c) $\frac{12}{25}$

PRACTICE EXAM PAPER

Paper 1F

1 (a) (i) 7°C (ii) −11°C
 (b) (i) 6°C (ii) 10°C

2
Date	Deposit	Withdrawal (£)	Balance (£)
01/05/15			4240.00
06/05/15		300.00	3940.00
15/05/15	345.00		4285.00
19/05/15		450.00	3835.00
27/05/15	1350.00		5185.00

Kate does not have enough money for the garden patio.

3 (a) 2 (b) 0.5
 (c) Same chance because the probabilities are equal
4 (a) 14 km
 (b) Harry because he walks 39 km but Lewis walks 37 km
5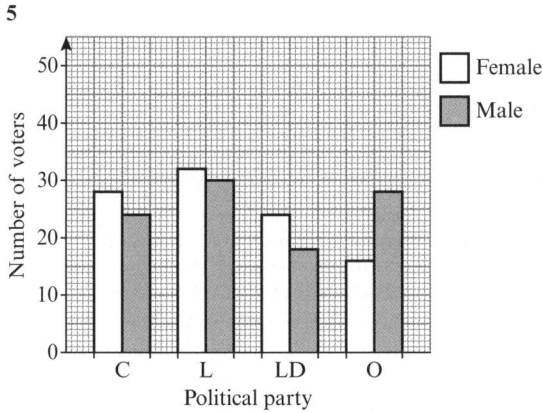

6 (a) $4a$ (b) $10x$ (c) $19e - 8f + 4$
7 (a) $25 - (10 \times 2) = 25 - 20 = 5$ Len is correct
 (b) 16 (c) $16 - (4^2 + 3) = -3$
8 $\frac{3}{4}$
9 (a) Angle ABC is 60° (vertically opposite angles are equal) and angle ACB is 60° (angles in a triangle add up to 180°). Hence, all the angles in the triangle are 60°.
 (b) 110° because angles on a straight line add up to 180° and angle SQR is 70° (since base angles of an isoceles triangle are equal and angles in a triangle add up to 180°).
10 780 m
11 (a) $5(x + 2)$ (b) $x(x - 6)$
12 (a)

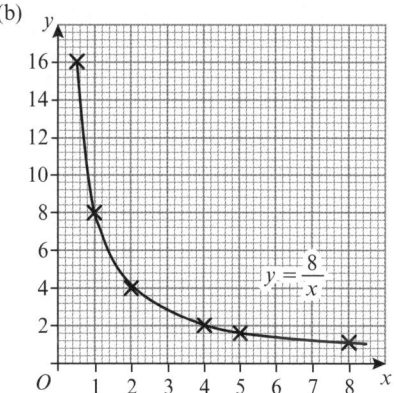

(b) £670

13 11 glasses
14 £11 475 < £11 500
15 (a)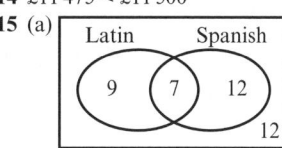

(b) $\frac{12}{40} \times 100 = 30\%$
16 9
17 16
18 nth term is $4n + 3$
 $(4 \times 31) + 3 = 127$, so 127 is a term in the sequence.
19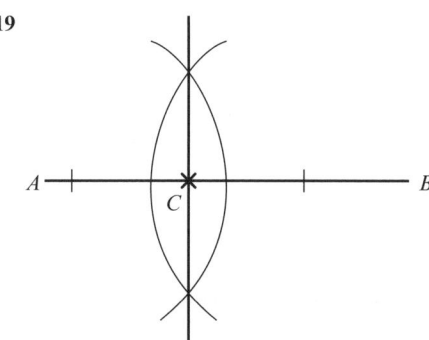

20 (a) (i) 5×10^4 (ii) 0.000 096
 (b) 1.5×10^{11}
21 (a) 120° (b) 10 cm (c) 27 cm
22 (a)
x	0.5	1	2	4	5	8
y	16	8	4	2	1.6	1

(b) graph of $y = \frac{8}{x}$

Published by Pearson Education Limited, 80 Strand, London, WC2R 0RL.

www.pearsonschoolsandfecolleges.co.uk

Copies of official specifications for all Pearson qualifications may be found on the website: qualifications.pearson.com

Text and illustrations © Pearson Education Limited 2016

Typeset and illustrations by Newgen KnowledgeWorks (P) Ltd, Chennai, India and Tech-Set Ltd, Gateshead

Produced by Out of House Publishing

Cover illustration by Kamae Design Ltd

The right of Navtej Marwaha to be identified as author of this work has been asserted by him in accordance with the Copyright, Designs and Patents Act 1988.

First published 2016

23
19

British Library Cataloguing in Publication Data
A catalogue record for this book is available from the British Library

ISBN 978 1 447 98792 5

Copyright notice
All rights reserved. No part of this publication may be reproduced in any form or by any means (including photocopying or storing it in any medium by electronic means and whether or not transiently or incidentally to some other use of this publication) without the written permission of the copyright owner, except in accordance with the provisions of the Copyright, Designs and Patents Act 1988 or under the terms of a licence issued by the Copyright Licensing Agency, 5th Floor, Shackleton House, Hay's Galleria, 4 Battle Bridge Lane, London, SE1 2HX (www.cla.co.uk). Applications for the copyright owner's written permission should be addressed to the publisher.

Printed in Great Britain by Bell and Bain Ltd, Glasgow

Notes from the publisher

1. While the publishers have made every attempt to ensure that advice on the qualification and its assessment is accurate, the official specification and associated assessment guidance materials are the only authoritative source of information and should always be referred to for definitive guidance.

Pearson examiners have not contributed to any sections in this resource relevant to examination papers for which they have responsibility.

2. Pearson has robust editorial processes, including answer and fact checks, to ensure the accuracy of the content in this publication, and every effort is made to ensure this publication is free of errors. We are, however, only human, and occasionally errors do occur. Pearson is not liable for any misunderstandings that arise as a result of errors in this publication, but it is our priority to ensure that the content is accurate. If you spot an error, please do contact us at resourcescorrections@pearson.com so we can make sure it is corrected.